BIOPOL

Edited by Larissa Pschetz, ...
Illustrated by Pilar Garcia de Leaniz

TALES OF URBAN BIOLOGY

Biopolis: Tales of Urban Biology
Edited by Larissa Pschetz, Jane McKie, and Elise Cachat
Illustrated by Pilar Garcia de Leaniz

ISBN: 978-1-8381268-0-3

Published by
Shoreline of Infinity / The New Curiosity Shop
Edinburgh, Scotland

Design by Marco Scerri

A catalogue record for this book is available from the British Library.

If you enjoyed this book, find out more about what we do at
www.shorelineofinfinity.com

This book was supported by the Partnership fund between the University of Edinburgh and University of Sydney, and by the Edinburgh Futures Institute Research Award.

160221

BIOPOLIS

TALES OF URBAN BIOLOGY

The urban environment has long been ripe for creative proposals and debate. Planners, thinkers, and artists have questioned, praised, and fantasised about the qualities of city living, looking for ways to redefine practices and spaces. Practical proposals ranging from modern architecture to models of the 'smart city' have attempted to tackle issues emerging from urbanisation and population growth, while also permeating imaginaries of what cities could be like in the future. From reinforced concrete to sensor-based tracking mechanisms, new technological developments, combined with ingenuity and creative vision, have permeated assumptions of epoch, lifestyle, and conviviality, which in turn contributed to redefine cities across the globe. It is this interplay of dreaming, representing, planning and acting that prefigures the acceptance of designed elements as natural features of the urban environment, giving an air of normality to innovations such as skyscrapers, near-field payment technologies, and GPS-based tracking mechanisms.

The technologies that have transformed urban life in the last decades have been led in large part by innovation in physics and chemistry. Today, however, many are starting to see biological research as key to innovation in the future. New bio applications, products and services are emerging, potentially changing our lives in radical ways. Bio-based sensors, neurological computers, materials produced by gene-edited organisms, and new practices and roles associated with these technologies may one day become taken-for-granted elements of the cities we live in.

In this book, we have explored new visions of city living through the lenses of biological research. We invited a selection of prominent writers living in Scotland to engage in conversations with cutting-edge researchers based at the University of Edinburgh in order to create speculative stories about the impact of biotechnology on urban life. The resulting tales of brain simulation, mobile and malleable buildings, bio-recycling, biodesigned species, altered bodies and microbial communication simultaneously speak to our very human longings for connection, for recognition, for freedom; ultimately, they articulate our perennial aspiration to find better ways of living. Mostly set in Scotland, these are deeply human and universal accounts of bio-based futures that illuminate the relevance of research in synthetic biology, biochemistry, bioengineering, computational biology, and biomedicine to our collective goals.

We hope you will not only relish the stories but also gain new perspectives on what urban technologies and the spaces they are housed in could be like. 2020 has challenged many of our assumptions of shared and communal living. New bio-related discoveries offer tangible and enticing hints of alternative futures that allow for more informed and nuanced engagement with non-human actors as well as new perspectives on wellbeing. These are radical bio-related imaginaries. Enjoy.

Larissa Pschetz, Jane McKie, and Elise Cachat

Jane Alexander

A Temporary Structure

It was never an option, until it was. Even then it was hard for Sam to admit—first to herself, then to James—how ferociously she wanted what she'd always known she couldn't have. But once she'd voiced the question, it took them just one evening to decide. Five hours, two bottles of red, and a conversation that led them directly into the quicksand they habitually avoided by mutual, unspoken consent. Sam's mum. The disease. Sam's own prognosis. How long they might have together, as a family.

There were more practical issues, too: the technology was there, but it was still expensive; last year Sam had turned forty; their third floor, gable-end tenement flat—kitchen-diner, lounge and a single bedroom—was already cramped. None of it would be straightforward. But the sudden possibility, coinciding with her last years of fertility—surely it was a sign?

In the end it was the lack of space, not finance or fertility, that proved the most persistent barrier. Hopes captured by visions of a miracle grandchild, Sam's dad and James's parents were only too happy to contribute to the cost of the procedure. Of Sam's body offered up to injections and sedations, egg collection and embryo insertion.

A dozen eggs.

Two gene-edited embryos, both female.

One that sticks.

Less than ten per cent success in women over 40: they know how lucky they are.

Lucky. Lucky. Lucky. As she leans over the toilet at work, quietly vomiting. As she freshens her breath with ultra-strong mints. As she packs in a steady supply of oatcakes, tiny mouthfuls to pacify her rebellious stomach, and fights the urgent desire for sleep with half the workday still to go. All the while, Sam is singing the luckiness song, silently to herself.

For the first three months they don't tell a soul, not even their financial backers, the grandparents-to-be. When Sam gets back from work she crashes out on the sofa or crawls straight into bed. Meanwhile James spends the evenings on property websites, making eyes at ground floor flats with two bedrooms and a garden. But family homes are in short supply, and demand so far outstrips availability that they'd have to leave the city to find anything they could afford. They are spoiled with their present location, so handy for parks and shops; and it turns out their address falls into the catchment for an excellent primary school, a feeder for one of the best secondaries in the city. Not to mention

James's parents are a short walk away—a mixed blessing now, but Sam can see how useful they'll be once the baby arrives.

'We'll have to stay here,' she says, 'and keep her in a drawer in the bedroom.'

Beside her on the sofa, James looks up from the laptop. 'For the first year, fine. After that?'

'When she turns fourteen, fifteen, you mean...'

James sighs. 'I suppose we could get a sofa bed for in here, and this could be our bedroom for a bit. They used to have whole families in these flats, didn't they? Maybe we're just over-privileged.'

'Mm, how unreasonable of us to aspire to a separate bedroom for our child and a scrap of outside space we can call our own.'

'There's a two-bed semi on a new-build estate, offers over 265. Early viewing recommended. It's a bit of a box and the garden's small and it'd be a monster commute, but maybe...'

Sam pulls a face. 'Marigold is not convinced.'

Their collection of possible names is a carousel that spins in front of them, each name a promise of a different girl, a subtly different future. They will know her when she arrives, but her middle name has been set from the start. Sam's mother was Marigold, and their daughter will be Marigold in the middle, just right for a summer baby. There's something that amuses them both about the old-fashioned name applied to the tiny creature growing inside her. Already, Marigold knows what she wants. 'Marigold's hungry,' Sam will say, 'hungry for chilli crisps,' and James will lever himself up off the sofa to do the baby's bidding.

James's property searches leave their tiny footprints all over the internet, and before long targeted ads transform all their devices into shop-windows for houses they can't afford, far-flung flats in poor catchments they could just about stretch to, builders specialising in home extensions.

'Brilliant,' says Sam when these last promotions start to pop up, 'let's extend from the third-floor into mid-air, it's the perfect solution.'

And then James sees the ad for GrowPods.

At the 12-week scan Sam is somehow shocked to see a baby floating inside her. There she is: Marigold. Ghostly white, reclined comfortably on her back. She's almost all head. Sam can identify the curve of a cheek, the fainter smudge of her torso. Imagines zooming in and in, inside the baby that's inside her, until she can see the ribboning twists of her DNA, the precise site where she has been fixed. Even after the birth she and James will have to take it on faith that the gene editing has been successful (though the clinicians have never suggested it might not). It will be Marigold's choice, when she turns 18, whether or not to be tested.

Sam stares at the curves of her baby's profile, and a qualified relief seeps through her. It should be safe, now, to let people know. First they tell Sam's dad, who performs some rapid blinking, clears his throat and announces three times that he couldn't be happier; then James's parents, who'd apparently guessed when Sam had declined both prosecco and wine at his mother's birthday dinner (and besides, says his mother, not meaning to be personal but she's never usually so busty). And it should be safe, now, to seriously consider how they're going to accommodate their new arrival.

Between them they've visited the GrowPod website so often the ads have become relentless, crowding out the buggies, the Pampers, the washable nappies. *Now you can grow your living space where traditional extensions are difficult or even impossible*, it promises - but neither James nor Sam quite understands how the construction is meant to work. *A process that harnesses natural nanotechnology*, says the site. *The innovative GrowPod construction process uses precision nano-machines to convert carbon dioxide in the air into materials such as hardened cement, brick, and glass- and wood-like substances. These nanomachines are completely natural, taken from living plant cells, and programmed using DNA instructions. The resulting solution can build engineered structures with exquisite, nano-scale precision:*

- *floors, walls and ceilings that can then be quickly and easily carpeted, plumbed and wired*
- *built-in hard furnishings*
- *even custom paint-like or 'wallpaper' films, for a flawless decorative finish*

All of this with minimal disruption and at affordable prices, delivered with our professional expertise and a ten-year guarantee.

There is a gallery of images to click through. The exterior face of a tower block is adorned with brightly coloured pods positioned asymmetrically on several different floors, portholes like single eyes in cheerful faces. A sandstone-look pod with a single large window extends at first floor level into the side return of a Victorian semi. A more curvaceous design protrudes from the facade of a two-storey 1950s development; it seems to perch on the balcony that overhangs a shabby street-level arcade of dry cleaner's, Ladbrokes, tanning salon.

'Are they real, or Photoshopped?' Sam enlarges the shots, peering closely.

James shrugs. 'I guess we'd have to put it at the back, if we got one? For planning, I'm thinking.'

'They don't look real.' She clicks out of the gallery, back to the page that says *prices start from just £30,000.* 'Start from,' she says. 'I know it says affordable, but ... I think I'd need to see one, first.'

Of friends to share their news with, Freddie and Ellen wouldn't have been near the top of the list, but neither Sam nor James knows anyone else with a GrowPod. *It's been FAR too long*, Sam writes in her email suggesting she and James drop round for a visit, and as she presses send she feels this is true. Throughout their college years Sam and Freddie had been close, had shared a flat with a group of other girls. Then Sam had moved away for a while (her wild years she called them now, if she spoke of that time at all), and when she returned Freddie and Ellen had settled out in the western suburbs, one kid already in tow.

'It is bloody miles,' says James as they sit in traffic, halfway into the hour-long drive. 'No wonder we never see them.' But of course, it's not just distance. For years Sam has shrunk away from friends who've embraced the domestic. Slipped congratulations cards into the post, then muted them on Facebook. So she's never seen Freddie's boys in the flesh, never visited their 1930s semi with its GrowPod extensions.

As they take off their shoes in the lobby Freddie exclaims at this fact, and Sam finds herself reaching for excuses to present along with her gifts of flowers and cake. The distance between them, their busy lives...

'Ah, it just happens, doesn't it?' says Ellen. 'Time just vanishes.'

Sam has always felt slightly intimidated by the precision of Ellen's sharp black bob, her polish the opposite of Freddie's chaotic warmth. But when she makes her announcement it's Ellen whose face lights into uncomplicated happiness, while Freddie's registers surprise; Ellen whose arms open wide to welcome her into the world of motherhood, while Freddie is still rearranging her expression to something more appropriately celebratory.

'So,' says Sam, 'we have a hidden agenda. Could we possibly take a look at your GrowPods?'

While the boys are in the garden kicking a football, the adults squeeze into each bedroom in turn. Freddie points out the features, explains the choices they made. 'We always knew we wanted two kids, so we had both pods done at the same time. It's not about saving money, particularly. In fact, cos we customised a few things it wasn't that much cheaper than the quotes we had for a traditional extension. But it's so much less disruptive. If we'd had builders in, they would have been trampling all over the garden—next-door's too. We'd have lost all of those plants, see?'

'And also,' says Ellen, 'there's Sylvia next door, she has a rose that's a memorial for her husband, and that would have gone as well.'

'But with this it's easy, they just put a tank in the little front garden, fed it through the house to the back. It was a bit of hassle at the start when they set it all up, wasn't it, but then we just basically ignored it for—how long did it take for ours, four or five months?'

Ellen nods. 'It just builds itself in the background. They monitor it, with cameras. Come and check it in person every so often, make sure everything's happening the way it should. And it's eco, that's the other thing.'

'Yeah, the carbon footprint is like, ten times less, something like that? That was a big thing for us.' Freddie turns to the eldest boy, who's appeared in the doorway. 'Alright Conor?'

'Why are you all in my room?' he says, wriggling past them.

'Just showing it off,' says Freddie. 'You like it, don't you? Like your bedroom?'

He grunts a yes, grabs a foam blaster from the cubby under his bed, and hares off into the garden to shoot his little brother.

It's not till Sam and Freddie are in the kitchen—filling the teapot, slicing the lemon cake she and James have brought—that Freddie stumbles towards her question.

'So, um, what changed? Like, with ... cos I always thought you couldn't have...'

It takes moment for Sam to register: a weary realisation that one of her oldest friends has forgotten, or never fully understood. To be fair, they'd probably both been down several pints of cheap lager when Sam had told her; same as they were the night Freddie had cried about her dying father (or was it her stepfather?).

'I mean, I never couldn't,' she says. 'It's just the risk.' She takes a breath, the words lining up efficiently, not used in a while but as familiar in her mouth as her own teeth, from so many conversations with men who might have been serious. 'You know I have this genetic disorder, it's what my mum died of, and it's a fifty fifty chance I pass it on if I have a child. Except now they can go in there,' she rests her hand on her belly, 'and edit the DNA. Get rid of the mutation that causes the disease.' This is the new part of the story. The happy ending.

Freddie stands by the counter, a knife in one hand and the packaging from the cake in the other. Her eyes look pink as she shakes her head, and instantly Sam forgives her.

'God, it's amazing, isn't it?' Freddie swoops her arms in an all-encompassing gesture: she means this, the corrected baby growing inside of Sam; and the boys she and Ellen carried in turn; and the pods that house them, and all of it, just ... all of it.

Amazing. That's what James keeps saying, as the architectural technician explains the construction process, talks them through their various options. They'd been thinking of a pod that opened off their bedroom—perhaps they could turn the window into a door?—but as soon as the consultant points out the problems Sam realises he's right: bedroom access might be handy at first, but opening off the public space will be much more practical in future years.

'But because your building's the last in the block, that gives us more possibilities,' he says, opening the cupboard at the end of the hall. 'Mind if I...?' He pushes aside a heap of towels, reaches right in to rap the back wall with his knuckles. It makes a solid sound. 'Here, you see. That's the end wall through there. Knock it through, and sure you sacrifice a bit of storage but you don't lose a window—in fact, you gain one.'

James is nodding. 'Planning permission, though... Would they let us, do you think?'

'Well ... the building's not architecturally significant, if you don't mind me saying. And the integrity of the whole block's already compromised by UPVC windows in a couple of the flats. So when it comes to planning—I know a couple of folk in there, they're reasonable guys, helpful guys,' he glances at Sam, corrects himself, 'helpful people. They're taking a favourable approach, these days. They know there's a crisis, prices are crazy, folks like yourselves can't afford the space you need.' Though *folks like yourselves* includes her, Sam feels extraneous in this conversation; can sense the consultant reminding himself to address his spiel to both of them, not just the man of the house. 'And the thing about GrowPods is, relative to your traditional extension, they're reversible. Easy to remove. I mean, it'll last you—it's not a temporary structure—but in terms of how it alters the building itself, it's pretty minimal.'

He talks of nutrients, and instructions, and nano-scale machinery, all in a matter-of-fact way that is on the one hand comforting in its steadiness, and on the other weirdly unsettling. While James nods and says *right* and *yes* and *amazing*, for Sam every piece of explanation prompts another uncertainty. Either James understands this all much more deeply than she does, or much more superficially—and she can't tell which. The questions queue up inside her, too many to ask. Does it really grow, like a plant or a creature, or is that just a name? How is it assembled—will it build itself? What exactly will it be made of? How can the solution convey instructions, and how do the nano-builder-things make sense of them?

Is this whole explanation a metaphor, and if so what's behind—beneath—inside it?

These tiny machines—the pods themselves—are they actually, sort of, alive?

Before he leaves, the technician suggests two models that are likely to meet with the approval of the local planning department: a rectangular pod with clean, straight lines and a more organic form with a curved roof, gently sloping sides and a perfectly circular window. All they have to do is agree on one or the other. James is traditional in his tastes, but Sam finds herself drawn to the rounded version. It has a friendly look. The interior walls will follow the exterior shape, no sharp edges for Marigold to hurt herself against; and built-in places for furniture—an alcove wardrobe, a recess to accommodate the cot and eventually a bed—means efficient use of space. And if they're going to grow an extension, they may as well have something that can't be built in the ordinary way.

In the end it's a compromise. The organic pod with a sandstone-look finish—off the peg, to keep costs down. Well, almost. The technician had warned them: the more you customise the design the longer it takes and the more costly it becomes; the DNA instructions need to be altered, which takes time and expertise. But when they search the standard wallpaper swatches there are sunflowers and roses, poppies and ferns, there is blossom on branches and generic, stylised blooms—but no marigolds.

Christmas. They spend the 25th with Sam's dad, Boxing Day with James's parents, and the rest of the holiday preparing for construction to begin. In the first week of the new year, work gets underway. With permission from the other residents, the tank of solution is winched into the back green. The feeder line is pulled up the outside of the tenement, slipped in through the bathroom window and passed down the hallway. The installation team is very thorough; the line is taped firmly against the skirting so it's not a trip hazard. 'Wouldn't want you taking a fall,' one of the workers says to her, and the words *not in your condition* hover at the end of his sentence. With the storage cupboard demolished and an opening knocked through the gable, the construction tent is sealed to the external wall, allowing not even a millimetre gap for the outside air to get in. Sam watches the workers do this from inside the tent, sure-footed as they balance at the edge of the opening. It's a neat effect; like a seam stitched with the fabric inside-out before it's turned right-way-round. She is thinking of sewing because one of the things they will have to install in the traditional way is the window blind, and if her dad still has her mum's old sewing machine she intends to make it herself. Marigold-print fabric, of course.

Before the building process can start the tent will be sealed against the inside of the flat, too, and the solution from the feeder line vaporised to fill the air inside. There, in the hot dark, the nursery will grow, one protein at a time.

As she moves through the city, now, she notices them everywhere. When she takes a weekend stroll along the canal, they bulge from the tenements that back onto the towpath. When she drives to the supermarket past the high flats she notices a collection of tanks on the concrete expanse between the two north towers, then a series of feed lines snaking up into the construction tents—serviceable grey, discreetly branded with the GrowPod logo—suspended ten, fifteen, twenty storeys into the air. And then she comes home, and is greeted by the comforting sound of purposeful activity. Through the length of the flat, the feeder hums. It snakes from the bathroom along the hall, past the bedroom and into the tent, delivering everything that's needed with the constant sound of a fridge or a speaker on standby. She and James have lived in this flat for almost ten years; she knows its every inch, from the ancient wine stain on the living room carpet to the crack that disfigures the bedroom ceiling. But the space is changed now, the flat more than itself. It's a conduit through which the feeder line passes to deliver its nutrition, its instructions. It's a host for the new addition, its bulk supporting the structure as it assembles itself: not yet part of their home, but not separate; neither inside nor out.

The flat is also colder than ever—near the bathroom especially, where the sash window is permanently raised to admit the chill of winter and early spring along with the feeder line. The temperature increases, though, close to the gable wall. Inside the tent, a certain level of heat and humidity is necessary for construction to proceed. Sam likes to hover there, for as long as the persistent ache in her lower back will allow, standing so close her toes brush the seam where the tent is sealed to the floor. When it was first installed it smelled faintly of chlorine. That's worn off now, or else she's so used to the smell that it no longer

registers, not even when she rests her hands against the taut fabric to enjoy its warmth, or presses her cheek to the rubberised surface—enjoying the friction that seems to hold her to it, the slight vibrations that might be caused by the feeder line entering the tent, or perhaps by tiny movements within. By the creation, from almost nothing, of something new.

The hum of the feeder is audible from every corner of the flat, steady and low, at the edge of hearing. But it's most noticeable in the bedroom, when James is asleep and Sam lies wakeful. When the only other noise is James's slow, sleep-deep breathing, she is able to tune in to subtle variations of pitch and tone, and wonder what they signify. Perhaps, from time to time, the nursery needs something different—more nutrients, a change of formula in the vapour—and what she can hear is the feeder line's response to that demand.

She's in bed when she feels it first: past midnight and she's still awake, curled on her side, afloat on the wash of sound. It dips, holds a note that plays in her head as the opening to a lullaby. *Twinkle, twinkle, little star*... And honest to God, that's when it happens. An extraordinary sensation, like bubbles popping inside her. It's so unexpected she catches her breath—and if James were not so fast asleep she would shake him awake, share the excitement, but instead she lies still as can be. Cocooned in sound, she focuses inward, alert for another flutter. For the tilt of a head. The open-and-close of a tiny, growing hand.

At the 20-week scan Marigold is so lively the sonographer struggles to get a clear image, but her heart beats clear as anything. Soon James can feel her move, resting his palm on Sam's bump to feel the jab of a fist, an elbow or a foot. They learn her habits: how loud music draws out all her angles, and spicy foods make her wriggle. How she likes to dance when Sam's trying to sleep, forcing her out of bed for a pee several times in the night. Each time Sam emerges shivering from the bathroom she is drawn past the bedroom door to press herself—bump, cheek, ear—against the tent. Sometimes as she hugs it close, absorbing the warmth, she will recognise the note it's playing. *Twinkle, twinkle*... Softly, she will hum along. The repeating pattern, ascent and descent. The return, at the end, to where the tune began.

They've been batting names back and forth again, their favourite dinner table talk. James is clearing the dinner plates, Sam rousing herself for the move through to the living room. 'Not Emily,' she says, 'there was an Emily at my school, and I hated her—' and her words sound too loud, in an unexpected silence. She's aware, suddenly, of the stark space of the kitchen, of its corners and hard surfaces.

Their eyes meet, in a mutual question.

'Is it finished?' says James.

'Can't be, it's way too early. Isn't it? It's run out of juice, perhaps—needs more solution?'

James frowns, pads through to the hall. When Sam follows, he's crouched on the floor, palm pressed to the feeder line like he's checking for vitals. He looks up, shakes his head. 'Nothing coming through, I don't think.'

Sam edges past him, stops by the soft grey wall of the tent. Tilts her head, as if she might pick up some kind of message from within. 'Can't hear anything,' she says. Tentatively, she reaches out a hand. The fabric is warm, but still.

They stay for a moment, undecided: him crouching, her standing. But it's too late, tonight, to call the technician. They'll phone in the morning, first thing. In any case, James suggests as they move through to the living room, it might be nothing at all. Might fix itself overnight.

They're in the habit of watching TV with the volume loud enough to drown the persistent hum; now when they settle on the sofa they keep it turned up to cover the absence of sound. Later, in bed, James drops off easily as he always does, but Sam lies awake. It feels like she's missing something: a blanket, or a pillow. The weight of the baby is pressing against her bladder, but the flat is even colder than usual and she puts off going for a pee as long as she can. When she finally gives in she keeps the lights off, follows the feeder line into the bathroom, its thick skin inert against her feet when she rubs against it. When she's done she hurries straight back to bed, huddles into James's warmth and pulls the duvet tight around her.

The silence is normal, she tells herself. Just a pause, a settling. A re-gathering of resource, before the work of growth can start again. The absence of sound doesn't mean, doesn't have to mean, that something's wrong.

The sonographer's nails are painted with a translucent silver glaze, subtle against her dark skin and flawlessly done. Sam wonders if it would look half as good against her own pale fingers. After the baby, she might try it. Not long to go now. She's sure the risk from the chemicals is exaggerated, but isn't it better to be safe?

All being well, this will be the last scan. She links her fingers, rests her hands on her chest. The gel on her belly is warming fast. She angles her head towards the screen, hungry to see the hot little star in the dark of her belly, glowing bright in the inner space.

A three-quarter profile this time: the chubby slope of her cheek, faintest blip of a nose. Left arm raised—fingers and thumb, count them, yes yes, just as they should be—like she's frozen in a stretch, like she's reaching for something.

It is such a clear picture.

The image shrinks as the sonographer splits the screen; the bottom half ready, like last time, to track the baby's heart in sharp white peaks, in a series of quick beats like persistent knocking at a door: *I am here, I am waiting, I am here.*

Close at her side, James draws in a breath, and holds it. It's the force of her will that stops his mouth—because as long as he doesn't say the words, as long as the question isn't asked—

'Is there something wrong?'

The technician is frowning over his tablet, ignoring James's question. 'This almost never happens.' He presses his lips into a thin, serious line. Taps, and swipes, and angles the screen to show them a series of shots from the cameras inside the tent. 'What it looks like is a corruption of the instructions. See here,' he says, 'the angle of that wall. That's a malformation there. And I know you specified a custom pattern for the decoration but,'—he pinches, zooms, and his face crumples—'*that's* surely not what you wanted.'

She can't make it out. The wrongness. His hand is in the way, and the light's bouncing off the screen.

He nods. 'DNA corruption. I'd say that's what it is.'

'But is it serious?' says James. 'I mean, is it ... can it be fixed?'

The technician stares at the tablet, like he can't look away from the disaster it shows him. 'Almost never happens,' he says.

The sonographer moves the transducer. Presses down on Sam's belly. Leans into the screen—and her head, now, is in the way, so Sam can't see, and the machine-hum fills up her ears, and she strains to hear the sonographer speak but no she's not speaking, her lips are tightly closed, and the moment stretches, stretches, and Sam holds her breath and waits.

To be told *everything's fine*. For Marigold to knock. For her small heart to say *I am here—I am waiting—let me in.*

A Note on the Science
by Nadanai Laohakunakorn

By breaking open living cells, scientists can capture their molecular machines—enzymes—and use them to carry out biological processes free from the constraints of cellular life. Central to these is protein synthesis, the key reaction that cells use to construct their components as well as their structure. The extracted biochemical soup of enzymes can be directed to build specific proteins of interest using instructions encoded in synthetic DNA. The study of the science and applications of this technology makes up the field of cell-free synthetic biology. While constructing an entire building using such a method is still science fiction, cell-free applications today include the rapid, low-cost production of vaccines and other therapeutic proteins, and paper-based biosensors for the detection of pathogens and environmental contaminants.

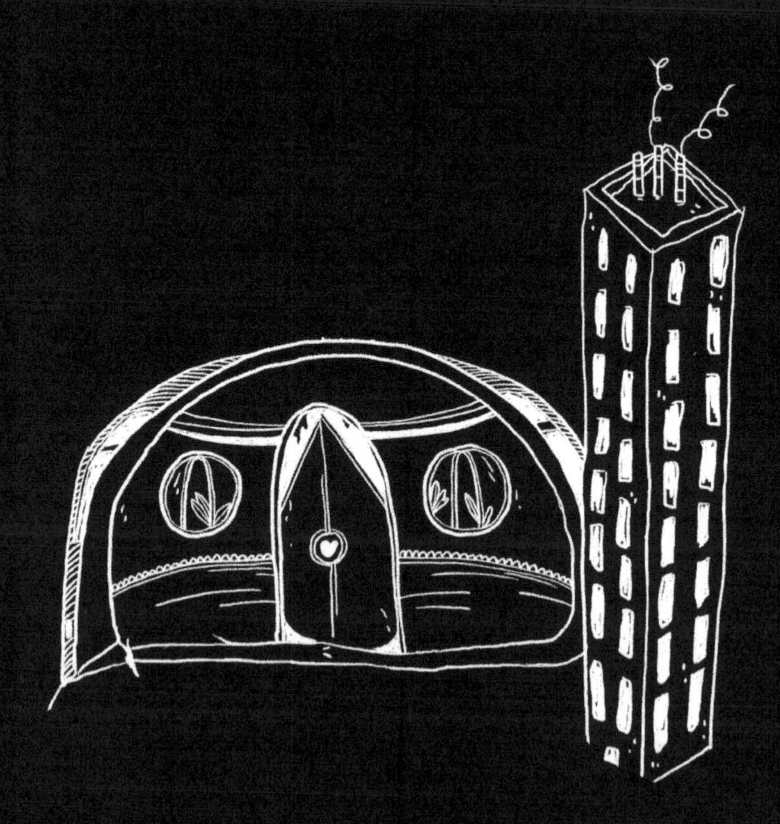

Jane Alexander is a Lecturer in Creative Writing at the University of Edinburgh. Her first novel *The Last Treasure Hunt* was a Waterstones Debut of the Year, and her follow-up *A User's Guide to Make-Believe* was selected as one of the cultural highlights of 2020 by the *Scotsman*. Her short fiction has won awards and been widely published, and a collection of uncanny short stories about science and technology is forthcoming in 2021. She holds a PhD in Creative Writing from Northumbria University.

Dr Nadanai Laohakunakorn, originally from Bangkok, obtained his BA/MSci in Natural Sciences (Physics) from Trinity College, Cambridge in 2010. He remained in Cambridge to carry out his PhD under the supervision of Prof. Ulrich Keyser, where he studied nanopores and single-molecule biophysics. In particular, his work focused on electrically-driven fluid flows generated within the confined geometries of nanopores, and he developed techniques to measure and characterise these effects using optical tweezers. After defending his thesis in 2015, he moved to Lausanne where he worked with Prof. Sebastian Maerkl at the École Polytechnique Fédérale de Lausanne, on the new and growing field of cell-free synthetic biology. His work there focused on combining microfluidic devices with cell-free gene expression systems for high-throughput and rapid prototyping of genetic parts and circuits. In 2019 he received a Chancellor's Fellowship in Biotechnology at the University of Edinburgh, where he has established a lab for engineering cell-free systems.

Andrew J. Wilson

The Negotiators

'Bacteria keep us from heaven and also put us there.'

—Martin H. Fischer

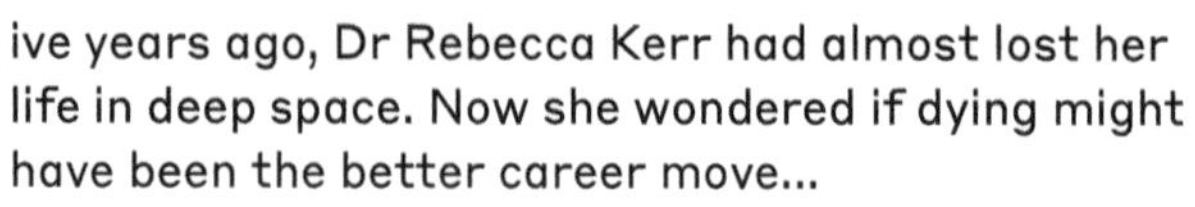

Five years ago, Dr Rebecca Kerr had almost lost her life in deep space. Now she wondered if dying might have been the better career move...

Her medical research had reached a dead end. What little funding the group had left was about to run out. The bacteria were still not cooperating.

'What do you want?' Rebecca asked.

The colony under the fluorescence microscope gave no indication of needing anything. Even though the microbes had been grown in a perfectly nutritious broth, the cells refused to divide. It was almost as if they knew that any increase in their numbers would only lead to trouble.

'You know what we want, Rebecca.'

She looked up from the monitor to find that Professor Gordon Boyle had entered the otherwise deserted laboratory. The head of the department had a face like a bunched fist, and the tone of his voice was just as ominous.

'Don't get up,' he said.

Rebecca ignored Boyle's suggestion, and the servomotors in her exoskeleton hummed in chorus as she rose. He flinched. Despite her atrophied muscles and emaciated frame, the aluminium structure around her intimidated some people. It was like a suit of space-age armour, and the built-in gyroscopes that compensated for her compromised sense of balance gave her movements an eerie grace.

'Can I help you, Gordon?'

'I need to hear some good news, Rebecca,' he said. 'I can't throw you a lifeline without one.'

That's a curious turn of phrase to use, she thought, *when you're one of the sharks circling this project...*

'What would you say if I reminded you that no news is supposed to be good news?'

'I would tell you that the road to hell is paved with good inventions,' Boyle replied.

'Intentions.'

'I beg your pardon?'

'The road to hell is paved with good intentions, Gordon, but I suppose you might find a few inventions along the way...'

Boyle glared at her. His annoyance at being corrected and then verbally outflanked boiled over.

'I need a progress report for the oversight committee,' he stuttered. 'Now!'

'At least give me a day to brainstorm with the rest of the team,' Rebecca said. 'As you well know, you can't rush good science, and you certainly can't ask an experiment to do what you want, pretty please...'

'Well, what do you expect me to do then?' Boyle asked.

'I expect you to wash your hands on your way out,' she replied.

*

Rebecca tried to relax as a tram carried her home across Edinburgh. Her stomach growled, and she decided that she wanted to order a pepperoni pizza. There was already a bottle of wine with her name on it in the house.

Tonight's dinner menu would be very different from what she had put up with as an astronaut. Those nutritionally balanced meals served in tubes and vacuum packs had been consigned to the recycling bin of her personal history. Rebecca never wanted to eat stuff like that ever again, but she would stomach anything if she could go back into space.

That wasn't possible, of course. Her last mission had made sure of that. It had been a desperate attempt to rescue a colony ship crippled by a meteor strike on its way to Mars. It had taken two years to complete the round trip, and radiation and weightlessness had taken a brutal toll on everyone involved.

As the sun set, the clouds glowed salmon pink and then rose red, reminding her of a heat shield on re-entry. When night finally fell, the plants lining the tram route began to glow in the dark. The bioluminescent bacteria that had been encouraged to colonise the trees and bushes on either side emitted a gentle azure light.

You do what the city wants you to do, Rebecca thought. *So why won't your cousins cooperate with our experiments?*

An idea as bright and sudden as a flash of lightning crossed her mind.

The exoskeleton buzzed as she sat up and fumbled in her bag for her mobile. The other passengers glanced at her to see what the commotion was, but she couldn't care less. She had made a breakthrough, and she'd be damned if she wasn't going to share it with her right-hand man.

Rebecca had Martin Ross on speed dial.

'Quorum sensing!' she said as soon as the group's lab manager picked up.

'Quorum sensing yourself!'

'Hear me out.'

'OK, Rebecca...'

'Molecular signalling regulates bioluminescence. Autoinducers alter gene expression in response to cell density. Bacteria can communicate with each other both within and between colonies...' Her stomach rumbled again. 'Can we communicate with them? Would they be able to talk to us?'

'I don't know,' Martin said, 'but I can't think of any reason why we shouldn't try. What have we got to lose?'

'Let's get the team together tomorrow morning.'

'See you then.'

The tram came to a stop, and the other passengers disembarked. An elderly man was the last to get o!f. He turned to Rebecca as he reached the doors.

'I couldn't help but overhear you,' the man said pleasantly, 'and it really is none of my business ... but could I make a suggestion?'

Rebecca smiled at him. 'Be my guest.'

'Make them an offer they can't refuse,' he said as he stepped onto the platform. 'Quid pro quorum sensing, if you will. Enjoy your pizza.'

The doors slid shut before Rebecca could reply.

She got off at the terminus. The stars came out as she walked home. Mars hung overhead like the last, dying ember of the sunset.

*

At ten o'clock the next morning, the team gathered in the departmental conference room. Sitting around the table were: Martin Ross, of course; Sandrine Lagarde and Jim McVeigh, the postdoctoral researchers; and Ravi Bal, a visiting Punjabi academic. They were all that were left of the group now. The PhD students had moved on after being awarded their degrees, and there was no money to fund any replacements.

The conference room was known as the Fish Tank because one wall was made of nothing but glass. Being in full view of any passers-by could be distracting, but it did allow for some advance warning if anyone attempted to gatecrash a meeting. Gordon Boyle, for example, had tried it more than once.

'Let's forget the agenda and get straight to the point,' Rebecca said, 'which is a particularly sore one for all of us... If this project is going to survive, we need to give our sources of funding something to show for it. Thoughts?'

'The gap we have to bridge lies between the micro- and the macroscopic,' Jim said in his soft Highland accent. 'If bacteria can mend breaks in their own double-stranded DNA—which they very de!initely can—why can't they be persuaded to do repairs for us?'

Sandrine was as elegantly blunt as ever: 'Why should they?'

Rebecca remembered what the curious old man on the tram had suggested: 'Why don't we make them an offer they can't refuse?'

'And how would we attempt that?' Sandrine asked. 'We already give them bed and board, so to speak.'

'We have to ask them,' Rebecca replied. 'We have to find a way to do it, and wait for an answer...'

Martin looked up from his notes and adjusted his spectacles: 'We could send them a signal.'

'Go on...' Rebecca said.

'Look, a lot of work has been done on quorum sensing,' he said, 'and we know that colonies can communicate remotely.' He tapped at his keyboard and brought up a schematic on the wall screen. 'Why don't we inject a synthetic genetic circuit?'

'Are you suggesting that we should encode an electronic microcircuit in their DNA?' Ravi asked.

Martin nodded: 'Nothing fancy, of course, just a binary-coded chip.'

'That would be a start,' Jim said.

'It would be better than nothing,' Rebecca agreed, 'which is what we have right now.'

'I can modify existing tech to fit the bill,' Martin said. 'I'll ask stores to send us what we need.'

'Who's going to volunteer to write the code?' Rebecca asked.

Sandrine and Jim both raised their hands.

'Good,' Rebecca said. 'Well, good enough. Let's get back to work. I'll give Boyle his damn report.'

Rebecca collected her notes while the rest of the team filed out of the Fish Tank, but Ravi hung back.

'You seem to take this very personally,' he said.

Rebecca nodded: 'I do.'

'May I ask why?'

'We risked our lives in deep space, and were rewarded with degenerative medical conditions. The thing is, my microbiota took as big a hit as my body. But—and it's a big but—they repaired themselves. They were the little bacteria that could. My body, on the other hand, was the big bag of bones that couldn't...'

She headed for the door.

'I took them all the way to Mars and brought them all back in one piece, Ravi. I think they owe me for that!'

*

On Saturday morning, Rebecca decided to head for the hills. A tram carried her to the southern edge of the city, and she took the path through the woods towards the high ground. She needed fresh air to clear her head and clarify her thoughts.

It was a bright day, and a brisk wind from the Firth of Forth swept what was left of the clouds away. The climb made the exoskeleton's servos sing their little song, but even with its support, her wasted muscles complained about the effort.

Rebecca ignored the pain. She was used to it, and she was happy to pay the price for some freedom. The Pentland Hills Regional Park was a tour de force of rewilding and conservation. A kaleidoscope of butterflies swarmed over the wildflowers on either side of the path. A honey buzzard looking for prey turned slowly overhead.

She reached the ridge, and headed towards the cairn at the summit of one of the hills. Out of habit, Rebecca added another stone to the conical pile when she reached it. Then she sat down in the lee of the little monument to eat her lunch.

There was a simple meal in her belt pouch. She had brought a cream cheese and walnut sandwich, a couple of Braeburn apples, and a bottle of water to wash down her medication. It was still better than anything she would have been able to eat during spaceflight.

The living city below her was now nearly as productive as the surrounding organic farmland. Almost every flat rooftop was a garden, and anything with a pitch was tiled with solar panels. Edinburgh's roads had been converted into green corridors for cyclists and pedestrians. The heavy traffic was carried along tram routes and railways driven by electricity from the offshore tidal power plants.

Human activity still generated heat, but none of it was wasted. It was ironic that spaceflight had taught people the lessons that they needed to learn. The struggle to survive offworld for any length of time had showed them how to live on Earth.

Rebecca's mobile pinged. It was a message from Martin. He had typed his message in all caps:

CONTACT—NOW WE'RE TALKING!!!!!

She wasn't sure whether it was the news or the fresh wind whipping around the hilltop that had taken her breath away. One thing was certain, the weekend was over and done with now. It was time to get back to work.

Rebecca walked over to the steep slope on the other side of the peak, and armed the trigger on her wrist. She pressed it as she leapt into thin air. Her backpack unfolded on cue, and spread itself out into a great pair of microlight wings. The smart fabric caught an updraught, and she soared into a sparkling sky.

Circling like the buzzards flying above her, Rebecca prepared to head for the lab. Still, she took a moment to savour her temporary release from gravity. She was made for this; she had been born to fly.

*

Boyle was not happy, and he made sure that Rebecca knew it. Even though it had only taken the group a week to get their first set of results, the professor didn't like what they were telling him. Now he had summoned her to his spartan office for a showdown.

'Are you serious?' he asked. 'Do you really expect me to believe that your experiments have gone on strike?'

'That's right,' Rebecca replied, 'and the rest of the group agree with me.'

'Have you all lost your minds?' Boyle gulped at his bottle of water and spluttered. 'That's absurd!'

Rebecca pushed a flash drive with copies of their findings across the desk: 'It's all here.'

After they had inserted the customised microcircuits into the bacterial colonies, the team had begun to introduce the microbial equivalents of carrots and sticks. The former were represented by various growth media, and the latter took the form of a broad range of antibiotics. The outcomes of each assay fine-tuned the next, and communication was tentatively established. As in computing, everything ultimately boiled down to binary code: yes or no; something or nothing.

'The key to this was establishing mutual terms of reference,' Rebecca said. 'It was easier than we thought it would be. It turns out we're not so different, after all...'

Just like computing, the questions that the team wanted to ask were converted into programming language for the microcircuits. In turn, the devices rendered this into zeros and ones, which the bacteria responded to with molecular signalling. It was a two-way process.

'Well, what do you think they want?' Boyle asked. 'Better pay and conditions?'

'As far as we can tell, they wanted to know what *we* wanted,' Rebecca replied. 'In existential terms, they were asking why they're here...'

'This is ridiculous!'

'Is it?'

Rebecca thumbed the catch that locked her exoskeleton in place, and surprised him by stepping out of the metal frame.

'We're not finished yet,' she said, leaning on the table for support, 'but it's a start!'

Boyle stood up to offer her a hand, but she waved him away.

'There's an old saying,' Rebecca said. 'Shy bairns get nae sweeties... I used our line of communication to explain that I wanted them to repair damage to human tissue.'

Boyle tried to step away from her, but banged into a filing cabinet in his haste.

'Did you inject yourself with one of your cultures? That's not just unethical—it's insane!'

'Calm down, Gordon, I didn't do anything of the sort. It would appear that they received the message loud and clear because they passed it on to my own microbiome...'

She sat down and beckoned Boyle to do the same.

'This is game-changing, Rebecca...'

His voice died away. Then Boyle went white and slumped in his chair. He twitched, rolled his eyes and sat up again with a jerk. It was as if he was an enormous puppet on invisible strings.

'Gordon? Are you all right?'

The head of the department smiled crookedly. It looked unnervingly like he'd never had the opportunity to try out the expression before.

'Never been better,' the puppet said, 'but Gordon's not available to take your call at the moment. You could say that he's on hold... How can we help?'

*

Rebecca gathered everyone in the Fish Tank for an emergency meeting. Even though the facts were hard to swallow, her team managed it. That being said, some of them looked as if they might now throw up.

'Boyle's eyes kept blinking every time I asked him a question,' she said. 'It was as if his brain was buffering—waiting for what he was supposed to say to load...'

Sandrine reached for her mobile: 'Have you called security, the authorities? Shouldn't we alert someone?'

'What would we tell them?' Rebecca asked. 'Something's possessed the head of our department, officer! Have you got any spare exorcists down the station?' She shook her head. 'This is our problem.'

'I agree,' Ravi said, 'but couldn't it be part of the solution too?'

'What do you mean?'

'You just have to join the dots, as it were. The bacterial colonies are communicating with each other in order to work with us.' Ravi waved his hand at her to underline his point. 'But this method of communication is too ambiguous and too slow. They know what we want, but they haven't been able to tell us what they're after, not until now. We need to discuss our terms...'

Rebecca nodded: 'Our gut feelings are the siren songs of our microbiota. They tell us what we want so that we eat what they need...' *Pepperoni pizza springs to mind*, she thought. 'We've just improved the lines of communication so that they can now get a word in edgeways too. It's a pity they've decided to turn Boyle into a puppet!'

Jim, who had been very quiet until now, slumped over the table, then lurched upright again. 'Not a puppet,' he said softly. 'The term *ventriloquist's dummy* would be more accurate...'

'Not you too!' Sandrine moaned.

Jim smiled reassuringly. The bacteria running the show were making a better job of it now than when they had taken over Boyle. Martin tensed and clutched his pen like a dagger.

'Please,' Jim said, 'calm down everybody. Don't be scared. We just want to talk.'

'That's easy for you to say,' Martin said, 'when you could seize control of any one of us next!'

'We're not taking anyone over. Just think about a driver and a passenger swapping places for a while. That's all this is.'

'Why didn't you possess *me* then?' Rebecca asked. 'After all, your friends are repairing the damage to my tissues.'

'Because we want to talk with you, Rebecca. We knew that Boyle wouldn't listen to us. We had to show him what we were doing by making an example of him.' Jim pointed at the glass wall. 'Here he is now.'

Their boss weaved down the corridor towards the Fish Tank. He looked drunk.

'Don't bother locking the door,' Jim said. 'If we needed any other representatives, we'd have co-opted them already. This is not a takeover. It's a negotiation.'

Boyle entered the room and sat down heavily in a spare chair.

'What do you want?' Rebecca asked.

'We want what you want,' Jim replied. 'We want to repair you and make you stronger. We want to make the long-term exploration of deep space possible, if not inevitable.'

Ravi looked him in the eye: 'Why?'

'We want you to colonise the solar system because we want to lay claim to it too.'

'We can only do it together,' Boyle said. 'O God! I could be bounded in a nutshell and count myself a king of infinite space...'

'*Hamlet*, Act 2, Scene 2,' Rebecca said. 'I didn't know the professor could quote Shakespeare.'

'He couldn't, but you can. It's your understanding that has changed our view of the world, and what lies beyond it. We all want to survive. We all want to thrive. We all want our space.'

Jim got up, switched on the wall screen and turned the channel to live drone footage of the surrounding area.

'Life depends on equilibrium,' he said. 'Humanity disrupted that, and your appetites nearly destroyed this planet. You learned your lesson just in time and saved the world. You had a little help from your friends, of course, but you did it. You restored the balance of nature by overcoming your own worst instincts.'

Jim changed channels to a feed from the spaceport in the far northwest of Sutherland. A rocket was being launched into orbit.

'However, you cannot stagnate,' Boyle said. 'If you are to survive and thrive, you must keep moving on, as you have done since your ancestors ventured across the Horn of Africa and into the unknown.'

'What if there's already life out there?' Sandrine asked.

'We'll cross that bridge when we come to it,' Jim replied, 'just as you've bridged the gap between us...'

Rebecca nodded.

'I believe you,' she said. 'Now, if you don't mind, I'd like you to make a gesture of goodwill.'

'We'd be delighted,' Jim said.

'Then please put our colleagues back in the driving seat. We have a great deal to discuss with them, preferably over a very stiff drink...'

'Whisky?' Boyle asked. 'A single malt, if we're not mistaken.'

'You're not,' Rebecca replied, 'but you're cheating. Not only can you read my mind, you lot put the idea there in the first place...'

*

Eventually, Rebecca only needed to wear the exoskeleton when she wanted to spread her smart wings. The microbes that she hosted within her body had overhauled it, and physiotherapy had done the rest. Her bones had been reinforced, and her musculature had been rebuilt. It hadn't been easy, but it was worth it.

She was the future: a human being synchronised with the environment that she lived in, and the one that lived inside her.

Now she sailed over the countryside almost every day. Taking to the air still helped her to think more clearly and deeply. What was more, the sky was no longer the limit. Her recovery meant that she could pass the medical that would allow her to return to spaceflight.

Her earbud pinged as she swept out over the Forth. She blinked to accept the call.

'Hello, Rebecca,' the bacteria said. 'How are you?'

'Oh, come on—you already know that!'

'We're only being polite.'

'Fair enough. I don't know what's more impressive—the fact that you learned manners from us, or that you can use our computer systems to hack into the mobile phone network.'

Inchcolm Abbey slipped past on the island below her. She banked north towards the coast of Fife.

'We're sorry to bother you on your day off, Rebecca, but we have some questions.'

'Fire away.'

'Which came first, the chicken or the egg?'

'Are you joking?'

'Why did the chicken cross the road?'

Concerned, Rebecca swung round again, heading back over the river towards home.

'Are you feeling all right?' she asked.

'We're trying to make a point,' the bacteria said. 'This whole episode is nothing but a horrendous paradox. You decided to communicate with us, but our preverbal selves may have planted that idea in your head. However, we could only have got that idea from you...'

Rebecca shivered as she formulated an answer: 'What do we have in common? Our source code—deoxyribonucleic acid.'

'Are you saying that we were all designed to do this?'

'I'm not sure,' she said. 'If we were, then that code was written a very long time ago...'

For a moment, Rebecca considered something that the Human Microbiome Project had once asserted: 'Microbes contribute more genes responsible for human survival than humans' own genes.'

Was it possible? Could the whole of evolution have been a bootstrap space programme? Was life on Earth designed from the start to be the vehicle to launch DNA into space? Only time, that most unreliable of narrators, would tell.

For now, Rebecca was happy to count herself a queen of infinite space again.

A Note on the Science

by Alessia Lepore

Understanding the strategies bacteria use to survive is important for human health: to fight bacteria that cause life-threatening diseases but also to preserve the bacterial communities that are essential to maintain human health, such as the ones living in our gut (gut microbiota). Scientists cannot talk or communicate with bacteria, but they use a variety of methods to study and understand them. Although they are unicellular organisms, bacteria are part of communities and need to communicate with their surroundings to guarantee their survival. Bacterial communities use quorum sensing to coordinate their population behaviour in response to changes in the environment: they produce signalling molecules that can be detected within their community but also by different bacteria. Some bacteria use quorum sensing to regulate processes such as the ability to emit light, also named bioluminescence.

All living forms, from unicellular organisms such as bacteria to complex multicellular organisms such as humans, have evolved strategies to proliferate and to survive the hazards they endure in their lifetimes. At the single-cell level, preserving DNA is crucial for survival as it contains all essential information that permits an organism to live. DNA can be damaged by different external sources, such as ionising radiations and UV light, but also by natural, error-prone cellular processes. Therefore, all organisms have evolved mechanisms to repair their DNA successfully. Scientists can genetically manipulate bacterial chromosomes to understand DNA repair mechanisms, and also gather more general insights into how the cell machinery works. The field of synthetic biology uses similar DNA manipulation techniques in order to create or re-design new biological parts, devices and systems for useful purposes, such as producing more affordable medical compounds. Using synthetic genetic circuits—cell parts assembled to perform logical operations as in electronic circuits—scientists aim to engineer 'biological machines' to perform useful, programmable tasks.

Andrew J. Wilson is a freelance writer and editor who lives in Edinburgh. His short stories, non-fiction and poems have appeared all over the world, sometimes in the most unlikely places. His stories have appeared in DAW Book's *Year's Best Horror Stories*, *Gathering the Bones*, *Professor Challenger: New Worlds*, *Lost Places* and *Dystopia Utopia Short Stories*. *The Terminal Zone*, his play about Rod Serling, has been re-staged several times, most recently under his direction at the 2014 World Science Fiction Convention in London. His poems have been published in *A Sea of Alone: Poems for Alfred Hitchcock*, *Umbrellas of Edinburgh: Poetry and Prose Inspired by Scotland's Capital City*, *Scotia Extremis: Poems from the Extremes of Scotland's Psyche* and *Multiverse: An International Anthology of Science Fiction Poetry*. With Neil Williamson, he co-edited *Nova Scotia: New Scottish Speculative Fiction*, which was nominated for a World Fantasy Award.

Dr Alessia Lepore is a postdoctoral research associate in the Institute of Cell Biology at the University of Edinburgh, Scotland. She grew up in Rome, Italy, where she studied Physics. During her PhD, she investigated how bacteria swim and how this motion could be exploited for technological applications. Then, driven by curiosity, Alessia joined the El Karoui Lab at the University of Edinburgh to understand the molecular processes that keep bacteria alive. She finds bacteria fascinating: such simple organisms and yet so evolved in their strategies to defend themselves from a variety of life threats. Alessia studies how bacteria repair their DNA after antibiotic treatments. She uses fluorescence microscopy to track the motion of proteins that identify and initiate the DNA repair process in *Escherichia coli*. Her findings help to clarify how bacteria repair DNA damage produced by antibiotics and to design more efficient strategies to fight pathogenic bacteria.

Gavin Inglis

Depth of Simulation

'There's a woman outside staring at you.' Jarrod's lunch companion gestured with his fork, as he chewed on liquorice-smoked lamb. Jarrod made a point of not looking.

People often peered through the windows of Epoch. The public waiting list for a table was eighteen months long, if you could afford its extraordinary service. Media figures could often secure something within a week. Jarrod had booked that morning.

Not that this guy was worth having lunch with: a young producer on a startup lifestyle channel. Jarrod's agent had set up the meeting. Jarrod wasn't enthusiastic, but she said, 'We're not thinking for now. We're thinking two years from now when you hit your late thirties and say to me, Cait, I'm done with streaming, I want something regular with a major. And I say, remember that guy you had lunch with two years ago? Well, he's the controller of UberLife now and he's been burning me with offers. Take the lunch. He's paying.'

Jarrod could tell the producer was new to Epoch because he chose the blood option. One little prick of the thumb by the *maître d'*, sixty seconds of multicore analysis and they had three courses tuned to your nutritional requirements and current mood. Jarrod always gave the breath sample. Less accurate, but also less hassle, and he didn't have to worry what they did with that blood later. He was satisfied with his depurple cauliflower fricassee served on celtuce leaves.

The producer sliced another juicy triangle of lamb. 'She's not going away,' he said. Jarrod frowned, and glanced around for a mirror. There was one behind the bar with an angle which offered him a three-quarters view of the unfortunate woman. She was unexceptional. Off-the-peg commuter jacket, shoulder-length disinterested hair, holding a hand at her brow to cut the glare from the window. He had expected a wide-eyed gawk at the culinary pleasures denied her, but all her attention was fixed on one point—presumably, his chair. Disconcerting.

Jarrod closed his eyes for a moment. 'I get so fed up of it,' he said. 'These people. They watch you at home all the time, so they feel like they know you.'

The producer speared fennel and radish with his fork. 'Don't look now,' he said. 'But I think she's coming in.'

Jarrod did look, a sideways glance without turning his head. The woman marched around to the restaurant entrance. He felt a ridiculous burst of anxiety. He had been over-reacting of late. There was that tearful moment a few streams back which had blown up his syndication by 79%. Cait had been delighted. 'You weren't faking,' she said. 'They love that shit. Cry more.'

Jarrod had a clear view of the woman as she pushed her way inside. It looked like she would sweep right into the dining room, but the *maître d'* was on her in seconds. It was standard now for these people to be enhanced, primarily for deference and flattery, but also with physical augmentations that made them excellent security. The woman clearly understood that, because she stopped just short of having her face pasted to the floor. She argued with the *maître d',* pointing towards Jarrod even as she was ushered back towards the entrance.

Their eyes met. 'Jarrod!' she yelled. She waved like a woman trying to stop a train.

The room looked at her, then at him. He looked at the cauliflower. Seconds passed, then there was a thump from the direction of the door.

'Relax,' the producer said. 'They gave her the bum's rush. She's hovering outside, but I don't think she's coming back in.'

Two minutes later, Jarrod spotted her, leaning on a router box across the road, watching the door. As dessert arrived, a creamy horchata semifreddo, the *maître d'* bent to whisper in Jarrod's ear. 'If you would prefer, sir, there is a discreet exit to the rear of the restaurant.' He felt absurdly grateful to the man. And once he had concluded things with the producer, expressing uncommitted enthusiasm for an ill-defined future collaboration, he headed to the back.

The transition between the front of the restaurant and the rear was abrupt. Out front, the walls were a worn olive with ebony detailing, vintage photographs of chefs and grand dishes constructing an imagined history of the establishment. Out of sight behind the ornate doors, processor modules formed the corridor along which these dishes travelled. Warm, gleaming bricks shaped the floor, walls and ceiling.

Even on a full stomach, Jarrod's nose twitched as he approached the kitchen. Their equipment was fully integrated, of course: heat from the processor cores rerouted to drive the ovens; load switching to power the rings on and off. He paused to soak it up: warmth on his face, the clatter of pans, the aromas; wondering also what was being cooked up around the room digitally. Endless simulations—financial, psychological, ecological—occupied the bulk of the city's compute power. The calculations for today's weather forecast could be distributed through the walls of a restaurant or a dance studio. Tomorrow, it might be a bank or a delicatessen. Jarrod lingered for a few more seconds until he noticed a white-jacketed chef waiting to get past with a gritted smile of murderous politeness. He cleared out through the back door.

Jarrod found himself in a warm alley lined with vents. Before emerging, he performed his secret transformation, the one he used every day to get around. He took his distinctive spray of hair and mussed it down over his eyes. He added a hunch to his shoulders and a slight lurch to his gait. Then he turned into the street and walked four blocks without anybody paying him the slightest attention. He could afford a chauffeur, of course. But comedy needed raw material, and that's what he got from the streets.

Empathic buskers were the new thing. These villains had enhanced their perception to the point where one good look at you was enough to know what music would spear you through the heart, would root you there and open your wallet. Jarrod spied a girl with a violin, swaying dark-eyed behind a jungle of blonde hair. He kept his eyes on the pavement.

The scent of fresh bread distracted him. A new baker had replaced Harrigan's; one of the hiptech indies which had pricing tiers linked to the level of simulation. They modelled microclimate, temperature gradients, knead geometry, rehydration and a hundred other variables. For the price of a beer, you could have a Superior loaf. For the price of a good shirt, you could have an Elite. He'd heard about another bakery on the West End whose top tier loaf cost as much as a car. Supposedly a sophisticated palette could tell the difference; definitely an enhanced one could. Jarrod couldn't believe it was worth the outlay. Perhaps he could get this stuff into the show later.

Close to the studio, he saw two guys not fighting over a parking space. He had to swallow a laugh. These pinheads dropped money on combat enhancements, imagining themselves ruling the streets. What actually happened was that each simulation recognised the other, ran a matrix of outcomes and informed the combatants of the likely, indecisive result and its accompanying injuries. Marketing compared it to the samurai battles of old, where the entire confrontation was played out in a stare, and one man walked away in defeat without either sword leaving its scabbard. The reality was more like two dogs yapping at each other from the limit of their leashes.

Jarrod swirled all this stuff around in his head. He'd usually have a draft script by this point of the walk, trying out various word orders to make the punchlines land harder. Something about empathic bakers, perhaps, making loaves you had to fall in love with. Or buskers in combat. But it wasn't coming easily. His head was... *fuzzy.* There was no other way to put it. It had been like that yesterday too. But it was worse today.

Normally his enhancement carried him, possible narrative lines forking ahead like track before a bullet train. Punchlines would hang there for a couple of seconds, to be grabbed or rejected. In the early days there had been long sleepless nights; unable to dial down the possibilities, he constructed shows through the small hours into the dawn. It had taken six weeks of mindfulness training and hypnosis to get it under control. But today, he was off his game. He'd just have to fake it. He turned the last corner towards the building which housed his studio.

The woman from the restaurant was waiting outside.

She hadn't followed him. She just knew where he worked. He didn't broadcast the location of the studio, but it wouldn't be that hard for a determined stalker to work it out. Window references perhaps, or geocoding in the footage. There had been fans before, shy creatures anxious for sixty seconds of his time and a selfie. But instinct told him this one would not be so easily appeased. Jarrod called security.

He waited while a towering woman in ballistic vest and bodycams emerged from the building. She had a half-metre height advantage and a polite but uncompromising manner. Once she got his restaurant stalker ten metres away from the door, Jarrod ditched the disguise and went for it. The woman saw him crossing the road.

'Jarrod! I need to talk to you!' She bobbed, darting from side to side in the face of an implacable human wall. But Jarrod made it inside to the cool air of the lobby, and took a breath.

Why was this bothering him so much? On a good day, he would have handled it himself, with the same effortless, breezy technique that took him through the live shows and saw off hecklers. Perhaps he should work from home; just for a while, until he felt better.

For some reason he didn't want to get into the lift and face the studio staff. There was a lounge round back, where guests waited for their chance to appear on a variety of other shows recorded in the building. The lounge was quiet, just one woman and a kid. He skulked in there.

In the hall, an engineer was maintaining a processor unit. The brick dulled as she slid it free, adjacent units flaring briefly as they redistributed the load. She kept pulling, drawing a length of optical cable from the wall. Even from a distance, Jarrod could see an ugly stain across its surface. The engineer clicked her tongue in irritation and began to treat the sheath with a chemical spray.

Jarrod turned his attention back to the waiting area. The kid was jumping now, launching from two feet flat on the floor, repeating the same movement over and over from the same spot. The mother just watched. He noticed a repair on her sleeve, clumsy home stitching, and that the kid was wearing a tatty jersey too big for him.

And then he knew why he was upset.

It was before his operation, when he was starting to deteriorate. A family emergency had meant he had to spend the day with his father. The old man was filming at one of the outposts, and he couldn't cancel the trip. So Jarrod came along in the back seat. Two hours with no escape hatch. The old man thought this was a good time to bring up his grades, which were dropping. He banged on about hard work and application while Jarrod fidgeted on the back seat and tried to explain he just couldn't remember things like he used to.

They passed through a village; not one of the agri-consortium settlements, just a line of tired houses on the side of the road. Jarrod stared at the people sitting out front, beaten by the dust from the car. Their clothes were dirty and ragged, dejection etched on their faces. Flies circled. It was the first time he'd seen poverty and its consequences.

'If you don't stick in at school, this is where you'll end up,' his father said. Jarrod felt an icy clamp on his bladder as he watched the village disappear into the haze behind them. It wasn't just the squalor. Nobody would listen to these people. Nobody cared what they thought.

Three weeks later he'd got the diagnosis. Jarrod didn't understand a lot of the details, but his parents flinched at the word *neurodegenerative,* and neither of them would look at him for a while. Then he remembered the village, and that was when he became too scared to get into the car in case they took him there.

The kid in the lounge finally misjudged a jump and fell on his face. His mother gathered him up as his mouth fell open and the wailing began. Jarrod dragged himself to the lift.

He put his public face on as he hit the studio, trading a recipe with the receptionist, listening to the engineer's latest failed nightclub conquest story. He studied the social media summaries and drank a cup of young matcha and none of it mattered, because when he got in front of the mic he had absolutely nothing to say.

They stuck it out for an hour and a half, while Jarrod tried random street observations, a touch of politics, even art critique. Nothing sparked. The engineer kept his eyes down. There had been cold starts before, but fifteen minutes was usually enough to fire up Jarrod's chain of wit. Not today. They looked at each other in silence. Then he gave the engineer the afternoon off and got security to let him out a fire exit.

On the way home he got the idea that the streets were humming.

As he approached the corner of his road, he paused, and told himself he was being paranoid. But he glanced around the building anyway. Sure enough, the woman was there, waiting outside the stair door. How did she know his address? Perhaps he ought to stop asking. He climbed over the back wall and tapped on an elderly neighbour's window. It took some fast talk to prevent her calling the police and to get him inside.

The daylight withdrew from his fourth-floor window as he stared at the wall. Cait called. He expected some kind of nag about the missed episode but she didn't even mention it. 'I've had a woman trying to get in touch with you all day,' she said. 'Something private, apparently. I usually block them, but this one seemed more insistent than desperate. I promised to pass on her ID, so, here it is. Your decision.'

'I'll think about it,' Jarrod said. 'Cait?'

'Yuh.'

'Have I peaked?'

After a couple of seconds of agonising silence, she laughed. 'You'll know when you've peaked, babe.'

'How will I know?'

'I'll drop you.' She hung up.

Fear chased him into the small hours. He tried meditative breathing, landscape visualisation, but he just couldn't let go. Jarrod lay in the dark, flopping from side to side beneath the duvet. In the end, he went back to the same old memory, the one that never let him down.

The consultation room was warm, and the doctor's beard reminded him of a bristly brush from their shed. 'You're going into hospital for an operation, Jarrod. We're going to fix the little problem which has made it hard for you to do well in school. Would you like that?'

Jarrod nodded, not entirely sure he did like the idea.

'You're very lucky, because until recently, we didn't know how to do this at all. But now we can. Would you like to see it on the screen?' He showed Jarrod a picture of a blue brain with lots of threads leading through it and down its neck. Some of the threads were glowing red. 'This is *your* brain, Jarrod. You remember when we did the scan? We're going to get your nerves to repair themselves using what we call *stem cells.* They're very, very small. We get them from your back, and reprogram them in our lab. Then they repair your brain.'

The doctor looked at him as if he expected Jarrod to do something. Jarrod nodded.

'Until recently it was hard to be sure what the stem cells would do. But with this new computing power we have, we can make a model of the entire brain—your brain, Jarrod. Do you make models at home?'

Jarrod had made a model at grandpa's insistence; a plastic aeroplane. He had glued the wings on backwards.

'Your brain is so complicated that a model of it would be huge. Bigger than this building. Bigger than the city, even. But we can build it inside the computer now, and that's what we did. We ran a *simulation* of what would happen in your brain. Many simulations.

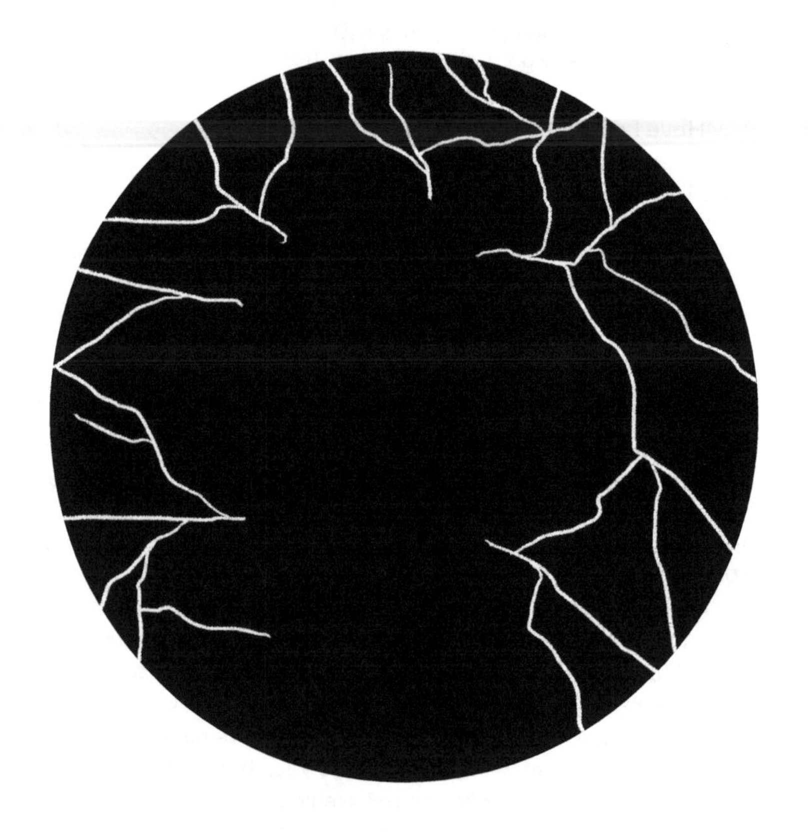

We found exactly the right way to use the stem cells to regenerate your nerves.' The doctor tapped his screen. The red lines faded to a cool blue and the doctor smiled. 'You'll be as good as new. Perhaps better.'

He could still remember that voice: confident, soothing, reassuring, those bristles jutting towards him. The voice that had made his parents relax. Jarrod descended into sleep.

Somebody shook him awake around dawn, and at first he thought it was the doctor. But it was a woman. 'Hey,' she said.

He scrambled back, holding the duvet up as a shield. Was he dreaming?

'I'm sorry to break into your place,' she said. 'But I tried everything else.'

He squinted at her. Yes, it was the woman from the restaurant. She was older than he thought. Her manner seemed calm; not the wide-eyed haplessness of the average fan. He needed to establish authority. 'How did you get in?' he snapped.

She shrugged. 'No software lock can resist enough processing power. I need to talk to you and it can't wait any longer.'

'Who are you?'

'Patricia. A medical engineer. I worked on your case back in the day.'

'My case?'

She sat on the bed. The intimacy startled him, but it didn't feel threatening. More like... a consultation. 'You had Cordwainer's disease. We thought it was MS at first, but you didn't show the physical symptoms. Just mood and cognition problems. Eventually they recognised the pattern of demyelination.'

'The pattern of what?'

'Imagine your nerves as electrical cables. They're insulated with fatty stuff called myelin. If the insulation breaks down, conduction is affected. Transmission speed. So you can't think as fast. It's quite a bit more complicated than that. But you just woke up.'

Jarrod blinked.

'You were one of the first patients where we used simulation data to program the stem cells. To decide how to treat you. You were our first big success, actually. The new oligodendrocytes produced excess myelin, and it went places that we didn't anticipate. You weren't just repaired, but *enhanced.* It bothered me that the simulation didn't predict it. But sometimes you have to take a win. When your show started, a group of us would get together to watch live. It was a nice team thing, you know? More than once I choked on my beer at your ad-libs.'

She peered between the curtains at the rising sun. 'You were a big help to the lab. We wrote papers about you, mentioned your show when we applied for funding. We still do, sometimes.'

He rubbed his shoulder. 'Why are you here?'

'Yeah.' She drew her legs up on the bed. 'This is what you need to understand about simulations. There's always a trade-off between the depth of your simulation and what's practical. With the processing power we have now, we can model your brain down to individual axons and dendrites. But that wasn't true when you were younger. We didn't want to regenerate you in a way which would become unstable later. So to complete the long-term simulations in enough time to help you, we had to simplify the model. Our simulation showed no complications, well beyond your life expectancy, so we went ahead with the treatment.'

Jarrod heard himself say, 'You simplified it too much.'

She looked down at her palms. 'We did what we could with the technology we had. But your enhancement was an extraordinarily good result. So, years later, I increased the depth of your simulation and re-ran it several times. I was trying to understand exactly what happened with the myelin. The last simulation was a hundred thousand times deeper. An... instability emerged in your middle thirties. They'll never confirm it officially, as that might admit liability—'

'You're here to tell me it's going wrong.'

Patricia clasped her hands and looked up at him. 'I'm afraid that you'll probably begin to experience some cognitive difficulties soon. Not unlike when you were younger. You may find it difficult to concentrate, or forget little things. When that does happen, I'd like you to—'

He threw the bedding to the side and stood up, startling her with his nudity. He grabbed his clothes and began to get dressed.

'What are you—?' she said.

'It's happening already. It's been happening for a few days. You're here to tell me that I need another procedure, right? So let's go. We'll do more scans so you can update your simulation. That's what you need, yes?'

After a few awkward seconds, she walked up to him, took the shoe from his hand and placed it on the floor. Then she wrapped her fingers around his.

'As we age, our body becomes less good at repairing itself. I've run the new simulations over and over. I'm sorry. There is no effective treatment at this stage. It's not even close.'

'But you don't have an up-to-date scan of my brain, right? So scan me again. I can pay for a more detailed—'

She held his hand tighter, and shook her head.

He pulled back his hand and sat on the bed, staring at the one shoe he was now wearing. She took the chair opposite.

'You came here to tell me I'm going to die?'

'Not imminently. But you will slow down, find it harder to concentrate. Eventually, it will be like it was before your operation, and then things are likely to get worse. I'm sorry.'

'But I can't... people will stop listening to me.' Saying it out loud made it real. 'I'll become irrelevant.'

'That's not so bad. People ignore me all the time.'

She wasn't referring to the restaurant. But he felt a spike of guilt anyway. 'You broke into my place to—just to warn me to prepare for this?'

'No.' Now she looked at her shoes. 'I have a request, and time is precious. I do want to scan you again. Once a week, if you can manage a regular slot.'

'Why?'

'This is still a young field of research, and your condition is very rare. We don't have much data on a longer timescale. The more we have, the better we'll understand what's happening to you and why.'

'So you can refine your simulations.'

'Right. Experimental data always tells us what the simulations can't. You could make a big difference for the next person with the same condition. Think of it as... think of it as staying relevant.'

Jarrod walked to the window, parted the curtains and stared out. The city stirred towards life, delivery vehicles slipping through the hesitant light of a new day. He got dressed in silence, and followed her out into the quiet streets. No need for a disguise, now. They passed unrecognised between glimmering towers, through open plazas. Some of these buildings could hold new pieces of him, even now; processing realities and possibilities, like his narrative line of forking tracks. It was a punchline, of sorts.

A Note on the Science
by Linus Schumacher

This story features several scientific ideas from my own work and that of colleagues. First, a core question in my own research is how stem cells interact to grow and repair tissues, as for example in neurodegenerative diseases. The particular disease here is made up, but was inspired by my colleague Prof. Anna William's research on multiple sclerosis. Then there is the idea that in mathematical models and simulations, we always have to simplify and leave out some details if we want to understand a complex system and make predictions. This is true in computational biology as it is in other areas of science, such as epidemiology. If an epidemiological model simulated the behaviour of every human being, as well as the air flow and the weather, it wouldn't be much use, because it would take as long to compute as it would to wait and see what happens in the real world. Last but not least there is the idea of ubiquitous computing, which ties into the 'city' theme of Biopolis. This is definitely still science fiction, but the idea of reusing waste heat from supercomputers is something I have heard being discussed for Edinburgh.

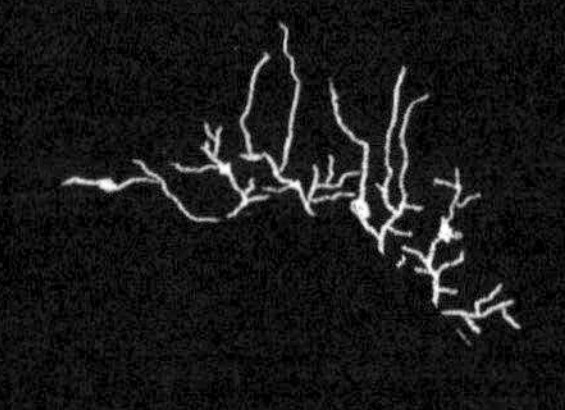

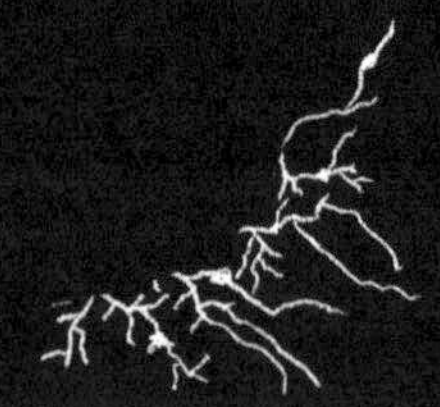

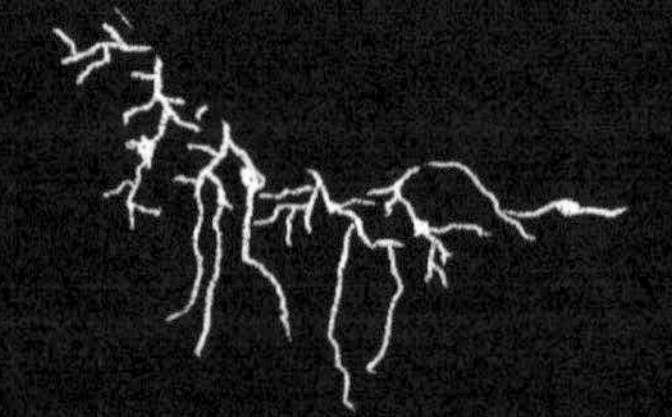

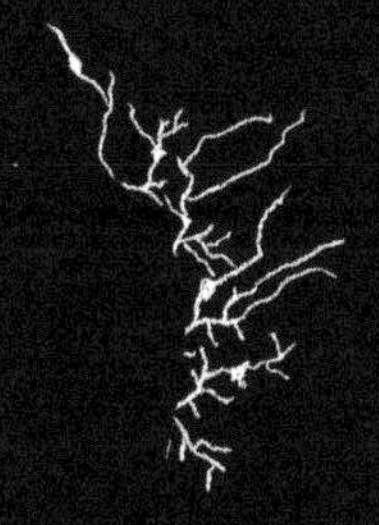

Gavin Inglis writes primarily for games, with interactive novels published through Choice of Games and credits on *Zombies, Run!*, *Call of Cthulhu*, and *Fallen London.* His interactive web story about mental health, *Hana Feels*, was nominated for an XYZZY award. Gavin is known for his compelling live spoken word, often with music. He performs with Edinburgh collective Writers' Bloc. Past collaborations include a city-wide poetry treasure hunt, a story set to smells, and spoken tracks with Glasgow electronic acts Spylab and Cinephile. Gavin was Language and Cognition Fellow at the NHS Department for Clinical Neuroscience, where he produced a graphic novel about Functional Neurological Disorder, AI versions of Jane Austen and H.P. Lovecraft, and a Book Festival show with live ASMR. He continues to straddle the digital/analogue divide to work on both old-fashioned 'turn to page XXX' gamebooks, and text generation by neural networks.

Based at the University of Edinburgh's Centre for Regenerative Medicine, **Dr Linus Schumacher** is an interdisciplinary scientist who combines biology, physics, and mathematics to study tissue regeneration at the cellular level. He leads a research group modelling interactions between stem cells (those potent basic pieces of life's make-up that can self-renew and, in some cases, repair damaged tissue) and the other cells that make up living tissue. Through mathematical modelling, the group explores what happens when cell groups interact and living tissue self-repairs to heal an injury or disease-related tissue damage.

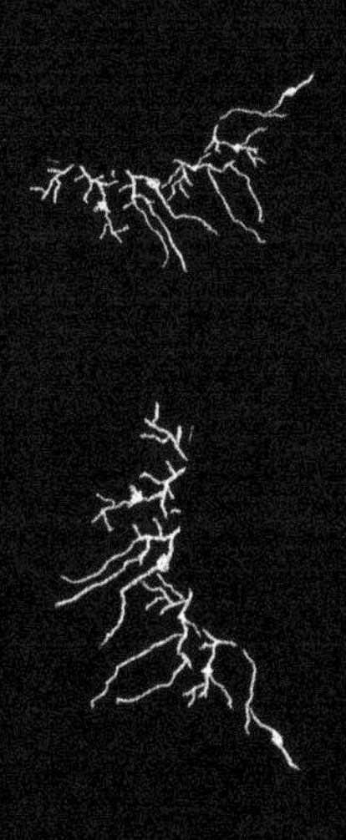

Alice Tarbuck

We Can No Longer Hold The Sun

Large-scale deployment of solar energy technologies will require half the current world supply of tellurium and 25% of the supply of indium. At the same time, the envisaged deployment of wind energy technology in Europe will require large amounts of neodymium and dysprosium (about 4% of the current global supply each) for permanent magnet generators, a demand that could only be eased if the supply of such metals in the future is increased, which may not be simple...

There were parades, when they reopened the mines. At first. Briefly, in all the depressed rural towns, there were prestige jobs. Mining companies recruited like the army: uniforms, pay, digs. Entertainment on weekends! Bonuses paid in data and smartphones, long after the hand-in amnesties started.

Our Brave Boys Below Ground, their faces bright on posters, on flickering televisions.

And the amnesties were gradual. Smartphones for certain professions only. Funeral homes meticulously extracting intelligent heart monitors. One computer per household, like something out of the 1980s. Then, one computer per household for certain demographics only. Off round your friends' houses to watch videos, to look something up. The end of self-drive vehicles that weren't public transport, everyone glumly back behind the wheel. The end of solar panels, which had clad the city like hot black slugs, trailing over every sun-facing surface. The end of personal windfarms, all landowners holding mock-funerals as the towers and processing plants were voluntarily laid to rest and repurposed.

*

Her wedding ring is made of solid Tungsten, encased in enough gold to make it skin-safe. Pre-reclamation gold, too, high carat and generously thick. Whispers say that it cost her fiancé more than three year's wages. They say he bought it for her on the blackest of black markets. Or that his clean, soft hands trade in something far darker. At that, they trail off significantly, raise an eyebrow.

Tungsten: the way it sits round her finger, so solid. Mocking, almost, when half of them now wear wood or scrap metal trapped in resin. Even that's pricey. Ribbon, they're offering, now? Stiffened, varnished, but still. Imagine.

Her wedding ring is made of Tungsten. He extracted it himself, he likes to say to dinner guests, smuggled it home in his pockets, ounces at a time, until he had enough. He tells the story rakishly, finishes it with a flourish, lifting her hand to be admired. This isn't true, but the weight on her finger is enough that she keeps her lips closed or smiles along with it.

She has a friend who lives in the mountains, where the air is vivid blue and clear, and they speak on the telephone when it is her turn. It is a party line, and they breathe into it, all matter of fact and coded. *The air's been lovely, the daisies are coming up, we've got fish and rice, I'm worried about the girls, up and down in that cable car. How's the city? How's business? No, no, I know. Take care.*

*

Of particular concern are 'critical' raw materials, so called because of their growing economic importance and high risk of supply shortage. The European Commission defines the following as critical: antimony, beryllium, cobalt, fluorspar, gallium, germanium, graphite, indium, magnesium, niobium, platinum group metals, rare earth elements, tantalum, tungsten

Her husband shows her the money, credits like a dragon's hoard, glinting on the screen of his expensive phone. Presses his dry mouth to her wet one—the money looks like food in all this heat, money like an aeroplane that will take them to another biodome. It makes her mouth water so much that he recoils. *Control yourself.* She controls herself, goes to the bathroom, closes her eyes, shudders silent as she masturbates. Runs the tap briefly, the water more aerated than wet.

Her husband shows her the money and tells her they will call their first child *Antimony*.

*

She is pregnant with her first child, and more than a little ill. What she has they don't have names for, yet, but they believe pregnancy makes the body more susceptible. They're trying to teach her not to dream of Trees. Not to hear them tapping behind the other sounds. *They talk a lot about roots*, she says, desperately. *They tell me about how empty the soil is.* Involuntarily, she finds herself in tears.

They are unsympathetic. Have faith in the flood of electricity they say. Watch the hanging lines that still dance. Put your energy into vertical gardening, into chemically separated soil, that black loam that expands like bentonite clay in water. Grow microgreens. Put your hand where your mouth is, but not empty this time. Put yourself where your mouth is. Step into your own mouth, it is the only way to fill it. Stop listening to things that have roots and no mouths. You know better than that.

That year, Antimony breaches the top ten baby names for the first time. Her Antimony, of course, is special.

Antimony, Cobalt and Graphite continue to rise in popularity, as this year's Top Baby Names list is dominated by Critical Raw Material inspired entries—but Birch, Elm and Oak aren't far behind as parents continue to celebrate our newfound 'entanglements' with our root-cousins. Scroll down to see more in baby names—new names, old favourites, and some surprising omissions!

*

They have one baby and it almost kills her. Afterwards, in the polished light of the hospital, the doctors make it very clear that she cannot carry another. The authorities provide a certification of risk, which means that the household will not be punished for their lack of productivity.

Once, in the night, she dreams another baby. In his cot, Antimony sleeps, solid and still.

The Court of Human Rights has convened today to discuss a response to the current raw materials shortfall and its effect on human health, nutrition and living conditions. Representatives from the Dendritic Council have petitioned to attend, but rumours from inside reveal that this request has not been accommodated.

*

There is a great deal of reporting around the first mine because it is, surely, an unlucky aberration. Simply something gone awry, an accident. There is a funeral procession, out on the streets, for twenty-seven miners. The streets are thick with mourners holding photographs of their smiling faces. By the fourth collapse, the state asks that individual biomes pay for the funerals out of their local budgets, so that by the seventh, there are no public funerals. In some cases, there is no way to get the bodies out: the equipment is simply too expensive. If the mine is to be sealed off, the bodies lie where they are.

The posters and the television campaigns do not quite ring true after the rash of collapses, after enforced early retirements, the illness, the mysterious pensions and deaths, the occasional discovery of unmarked graves. Eventually, the mines—still lucrative, still shiny and painted and full of perks—fail to recruit the base numbers needed. A global emergency is declared. Scarce things become impossible to source. It is a hot summer, even in the mountains, and things bubble, rise, and peak. She imagines her grandmother, newly married, unpacking new linens from their clean white cardboard packaging, learning recipes that might please. She imagines her grandmother would have been glad of a little revolution. She keeps that thought to herself.

A week into the biggest collapse so far, hopes for the fate of the miners have started to fade with rescuers expressing concern over the lack of oxygen inside the collapsed mining pit.

It is unclear how many are trapped inside, but the National Disaster Agency spokesperson said there could be as many as 100 miners still inside. This marks the inevitable closure of the last active fluorite mine in the northern hemisphere. The Rare Materials Commission warns of further shortages, and the non-negotiable withdrawal of consumer electronics for sale.

*

Antimony is healthy and strong. She rocks him and listens to the radio, in the shadowed afternoon, blinds halfway drawn. She doesn't want him dazzled, and the flat traps the sun. But then, she doesn't want him deprived of sunlight, either. He cannot go out until he is seven—or that's the latest advice. Over in the Silver Quarter, they take the babies out much younger, but the biome is different, there.

He's lucky, nevertheless. He might be stuck inside, but at least he isn't a mine baby. Mine-babies all have fresh-air names: Sunshine, Northerly, Hail, Snow. Pale things—bug pale. Born to burrow, to suck out what is left of the minerals in the dead soil by increasingly cunning and invasive means. Lots of nonsense in the press of course: that selective breeding over previous decades has made them blind as moles but able to smell precious minerals, sense them on the tongue, drawn towards them by the twitch of newly developed interior compasses. Why on earth should she believe that? They're just normal babies, delivered in mine-mouth complexes.

She donates to the relief causes, of course. Modestly, in line with her accountant's recommendations. Underground schools, proper sanitation. It's a difficult life, but those photos are mostly doctored. The press have said that. Sensationalists, attempting to create further outcry. The four-generation rule still stands.

Four generations ago, most of the miners' predecessors would have been resource-hoarding upper-classers who walked over the bent backs of families like hers, broken them in pursuit of more riches. And did. Her grandmother's hands, worn to stumps. Her blind grandfather, radiation readings so high he ended life alone, on an unstaffed ward, just AI and intravenous care. Suffering has a way of sharpening the mind toward justice.

And they'd had it all, that lot. Swanning in their air-conditioned vehicles, still taking commercial flights, taking holiday snaps and posting them online—online!—as if the servers weren't stretched taut, at breaking point. Vaults of precious metals, private nuclear reactors shared among the most privileged, and all the trappings of resource-wealth—hover cars, and re-specced mobiles, and those whirring, complicated prosthetics that could do things for you. Who could blame those who took over for wanting to breed the greed out of them? Hadn't it, after all, taken years to breed in? Requisition means freedom of the mind, not just the stripping of assets. Four generations of subterranean mining and, well, perhaps they'd emerge, blinking and sharp-nosed, with repentance in their supple, aching bones.

She turns her wedding ring on her finger. This is still—her accountant says—within reasonable personal ownership limits.

Not only are the Government-issue protein replacements excellent for almost all meat-recipes, they're also a fun stand-in for meals requiring nuts. The continued consumption of nuts, which have formed a cornerstone of our diets since the earliest times, are one of the urgent issues to be discussed at the inaugural Interspecies Conference later this month. Until a consensus has been reached, nuts and other Tree products have been largely suspended from sale

*

She finds, when the sun has moved its bleaching gaze away from the window, that her appetite returns a little. Cold water, stirred into bouillon, heated over the stove. She eats radishes with salt. She eats lettuce with a little oil. She eats powdered potato that the government sends in packets, reconstituted with oat milk. She is not well, and she is not unwell. Antimony eats their meat ration because young children are not given processed protein replacement until their adult teeth develop. Antimony swallows mince which comes from their cows, grazing sweet grass on the reclaimed industrial land below. She cooks it like her mother did: low and slowly, with an onion chopped so fine you could dissolve it. He eats it messy, because he is a baby. She wipes his face and holds him on her lap to watch the sunset. She slips a finger into his mouth, and feels how strong and sharp his teeth are coming in.

Over time, animal exposure to low levels of antimony causes eye irritation, hair loss, lung damage, and heart problems. Ingesting large amounts of antimony causes stomach pains and vomiting. Laboratory studies have shown that exposure to antimony can also affect fertility.

*

Her wedding ring is made of Tungsten. She glistens at the school gates. As if it is a virtue that she comes to collect the children herself, flouting government regulation—really it is only so they can see her in her glory. Perhaps the masks her family wear are better, they whisper. It wouldn't surprise anyone.

Solitary little Antimony, dragging his feet across the playground, as if hoping not to be associated with his fly-bright mother.

The rest of the parents trust the bus, its clean white emptiness, its filters. The doors whoosh open and suck the children in, its blind face sets off when the sensors indicate no further children are waiting. As a result, the smallest are often left behind to straggle home on their own, unless one of the kids on the bus realises, hits the button.

Barbaric, she says, her voice like wind over reeds. *Barbaric way to treat one's babies.*

Mum, Antimony says.

Still, little Antimony must hold her hand home, mask in place, parasol in place, march home across the softly needled ground. It is afternoon in the city, and the Trees are, for the most part, sound asleep. *Look*, she says, *listen to them breathing.*

Antimony wants to wake them up, talk about his day.

In science, he'd say, face against their pale bark, *we're doing your insides.*

None of the Trees in the park are linked up, so even if they're listening—and they aren't, in this heat they couldn't be—they cannot hear any reply.

Come along darling, she says, *before you take root yourself.*

*

When asked why they hadn't attempted to make interspecies contact with man before, the Dendritic Council simply stated that prior to this significant alteration in the earth's biome, it had not been considered necessary. Whilst the Trees had been concerned about man's relationship with the planet, the voting process had been slow. Messages from important Tree groups had to be counted and had taken years. Now, though, they would be staying more keenly in touch.

Antimony is playing with a sword made of driftwood. It was expensive, but they cannot police driftwood. He is banging the side of the cold-storage box, over and over again, and saying something that sounds like 'animal animal animal'.

A wooden sword is not, in fact, the same as 'making playthings out of human limbs found along canals', whatever the Trees might say. If they'd known Trees would be so much trouble when they woke, they might have put less effort into listening. But they've got a point about the bones, in a way. If it comes to it, she'll buy him human bones to play with. Lovely Antimony, his black curls bobbing round his face as he hits and hits and hits.

*

The afternoon that they break the door down, Antimony is churlish, won't talk to her or do his homework. She has a half hour on the phone to look forward to, later. Her husband will call. He is overseeing the demolition of some decommissioned factory and has been gone for weeks. The money is good.

The money is very good.

They will move, after this. Take Antimony and raise him in a different biosphere.

She's looked at the brochures of the more exclusive ones. Elephants. Beavers. Animals she's seen on film but he'll be able to see for himself...

The police find the metal deposits that are under the couch, the various ores that live inside shoes on the bottom of the shoe rack. *That was his plan*, she thinks: *enough to satisfy a raid, to stop them looking farther*. She cries, as she is supposed to. They don't take her in because there's nowhere to put her, no one looking after the boy, and she's not a threat. They leave her sweeping the mess with a dustpan and brush. They slam the door and tell her they'll send a locksmith round to re-secure it, since she's got a child. The Trees outside are giving statements. They tell the policemen that they don't feel safe with her in the neighbourhood. Imply she is an enemy of the soil. That night, when her husband calls, she cries down the phone.

A recent ruling from the high court concluded that testimony from Trees will, from next month, be permissible in court.

*

The only biome that will welcome them, after her husband has had a talk with the authorities, has flies. It has elephants too, apparently, but what she sees are flies. Their apartment has a balcony. It has marble floors in the bathroom, like the fancy hotels from vintage television programs about drug dealers, but even with two different widths of mesh, the flies come in, rise in blooms over the glass bowl of fruit, dance on Antimony's face when he's asleep. They buy flypaper and it twirls, sticky and black within minutes.

Her husband comes and goes, and she doesn't touch any new objects that appear. Here, a lot of the women wear metal rings. There is dense scrubland, the Trees few and far between. There isn't much water, either, in summer, but the well in the complex will see them through. Her husband says that underneath the sand, there are bound to be minerals, trace amounts too small to show up on government scans. For his birthday, she buys him a sieve, and a little, fine-needled detector. He works very hard, and she's so grateful. Antimony comes home from school himself, now. The bus has a driver; it is one of the old ones, refurbished.

Once a week, she phones her friend in the mountains. From the new biome, the reception is poorer, more sigh and crackle between the words. *We're seeing so many birds of prey, here, now. The shepherds are putting bells on the lambs. I'm sorry to hear about the flies, darling. When we've saved up enough credits—well. It would be nice to see you, see Antimony. See an elephant, perhaps.* The pips sound, and the line crackles into static.

Outside the apartment, the key turns in the lock. Her husband is home after weeks. Thirteen weeks. She had mended the holes in the sand-coloured tent herself, counted out the protein packets. She imagines him on the other side of the door before she sees him: thinner, tanned. His clothes will be covered by a film of yellow sand. He will have brought Antimony a fine, thin shell, from when the desert was all sea. She will think about all the things he has brought them when she wakes at night, listening for sirens.

Darling, he will say, *you'll never guess what I found.*

A Note on the Science
by Amanda Jarvis

The technological advancements of the 20th and 21st centuries have been vast. Our lives are now dominated by electronic devices—mobile phones, computers, televisions. By using chemical elements from across the periodic table, chemists have made materials that have driven the advances in electronics, screens, and batteries. These materials have also contributed to the development of new and sustainable energy sources such as solar power, batteries and wind power. However, whilst some elements such as carbon, oxygen and nitrogen are widely available, many elements are much rarer and/or harder to obtain. These include precious metals (e.g. gold, silver, platinum, rhodium), 'precious' because of their use as currency and jewellery across the ages, and the rare earth metals (the lanthanides, scandium, and yttrium). These metals are mined from the earth and are finite. If they are not recovered from products at the end of their lifetime, they are lost to humanity as tiny amounts spread across the world's surface in landfill sites. Due to increasing demand many of these metals are regarded as critical elements, which are considered at threat of running out in the next 100 years.

To maintain the resources needed for technology, scientists are investigating ways to recycle these metals, from developing molecules that will selectively encapsulate and remove one metal in the presence of another, to using plants for the remediation of metals from contaminated soils. Additionally, new materials are being developed that no longer rely on these critical elements to give them their unique properties, but use abundant elements such as iron, titanium, and carbon. Together these approaches should allow rare elements to be kept in circulation and reduce the reliance of technology on critical elements.

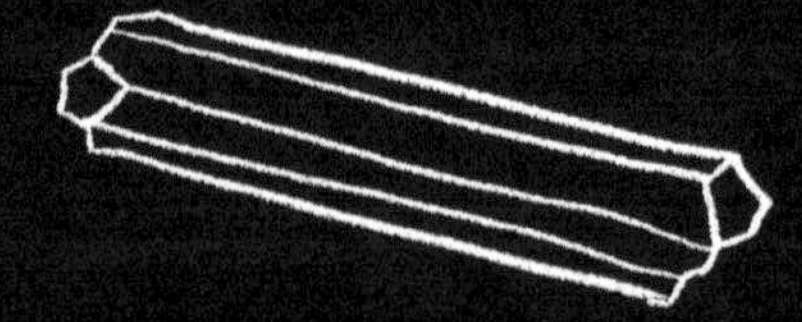

Alice Tarbuck is a poet living in Edinburgh. Her first pamphlet, *Grid*, was published by Sad Press in spring 2018. Her work has been commissioned by Durham Literary Festival, the Sheffield Post Office Gallery, the University of Edinburgh, Scottish PEN, and Timespan in Helmsdale. She was a 2019 Scottish Booktrust New Writers Awardee and is part of 12, an Edinburgh writing collective.

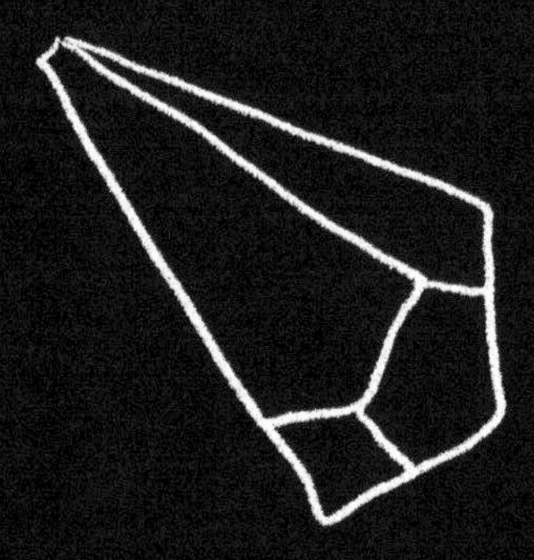

Dr Amanda Jarvis studies the design of artificial metalloenzymes, with the goal of using these new enzymes *in vivo* to produce cellular chemical 'factories' enabling the synthesis of complex products directly from sustainable resources. Dr Jarvis started her independent career at the University of Edinburgh as a Christina Miller Research Fellow in 2017, and in 2019 was awarded a UKRI Future Leaders Fellowship. She graduated in 2007 from the University of St. Andrews with a MChem and then went on to receive a PhD from the University of York. She then moved to France for her first postdoctoral fellowship at the ICSN and worked on the development of Rh(II)-catalysed nitrene reactions. In 2013, she moved to Professor Paul Kamer's group to work on sustainable catalysis, and subsequently received a Marie Curie Fellowship in 2015 to continue working in the Kamer group on Artificial Metalloenzymes for the Oxidation of Alkanes (ArtOxiZymes).

Kirsti Wishart

The Social Life of Cells

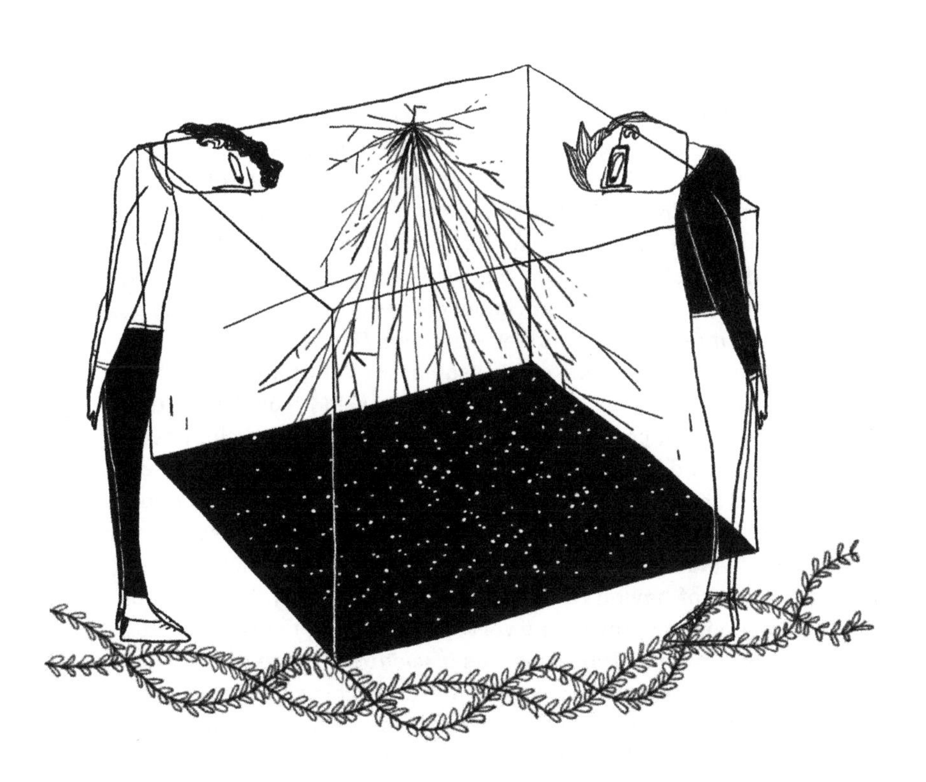

When the outbreak was over, the lockdown rescinded, Edinburgh experienced levels of Festival fever far exceeding those that usually met the beginning of August. Even cynics who made plans to escape the city during the weeks when you couldn't nip out for a sandwich without having to force your way through a Peruvian dance company, decided this summer they'd be staying put. When it was announced the first major event would be a revival of the Science Festival, the first Festival to be cancelled the previous year, the Council was overwhelmed with volunteers. As a mark of respect towards those who had successfully discovered a vaccine for the virus, Edinburgh would become a City of Scientists, a place to inspire new leaps in scientific thought. Ambitious science-inspired land art, sculptures and children's playparks and fairground rides would be constructed, the city desperate to have as many opportunities to gather together as possible.

When Fraser was given the assignment to meet up on the first day of the Festival with a biologist involved in the creation of some of the work on display, he very nearly turned it down. He didn't because the magazine that had asked him, emerging joyfully from the financially enforced slumber of the past few months, was a useful source of work. But one of the unexpected side effects of the slow release from lockdown was the effort it was taking him to adjust to being sociable again. Arts journalism had already occasionally felt an odd choice for an introvert, particularly when he'd find himself standing in the corner of some arty soiree. Then he'd wish he'd followed his original career choice of becoming a nun, announced aged five after watching *The Sound of Music*.

Months ago when one of his regular jobs that involved office work two days a week had switched to working from home, all interviews being conducted by Skype or Zoom or even just over the *phone*, he'd been struck by how much he'd preferred it. Yes, he missed some of his colleagues, the office chat, hearing about the latest documentary discovered on Netflix but he also secretly revelled in his solitariness. He knew he was being ridiculous but the thought of having to be in someone's company, their actual *physical company*, for at the very least an hour, nearly caused him a panic attack. Plus, this was a *scientist*. There'd be a load of technical jargon he'd have no clue about, his scientific learning having stalled after fainting during the dissection of a bull's eye during biology. He knew that was the editor's point, sticking them both together, showing how science should be easily translatable to even a numpty like him. Still, it felt like an interview to be endured rather than enjoyed.

As soon as he waved across to Anatol sitting in a booth in a bar on Lothian Road though, he relaxed. Anatol resembled one of the newer, trendier versions of Dr Who; tall and lanky but with a certain thrawnness too, retro-80s style wire frame glasses, a wispy goatee. There was an eager friendliness combined with geeky shyness that did away with any fears Fraser had about being intellectually intimidated.

They were taking part in one of the pop-up Eat and Drink Science! sessions appearing in some of the empty venues that hadn't survived the impact of the virus. As you drank your way through a beer flight—six glasses of a third of a pint—or ate your meat-replacement burger, an enthusiastic host would explain the chemical processes that resulted in an ale rather than a lager. Or you could debate whether the burger you were eating that had been 'grown' using animal cells was actually meat or not. There were coasters giving information on the likes of Angelina Fanny Hesse who had revolutionised the development of microbiology by suggesting the use of agar-agar in Petri dishes or advertising 'Anna's Brew', a beer honouring Anna Macleod, the world's first female Professor in Brewing and Biochemistry. Initially sceptical, Fraser began to enjoy the feeling of learning while slowly getting wasted.

'This is such a great idea!' Anatol enthused. 'People need to realise how natural a part of life science is, how it is all around them. It should be the case that labs are open to the public more, that people can come and chat to the scientists rather than us being locked up in buildings out of town. They should be like art galleries or libraries, where people can wander in but not be afraid to mix and chat. You know, at Summerhall, there is ACSUS, a lab where people, scientists and artists can go to, and I think more people should know about this. Because it is good for scientists too, we need to mix with other people, to be playful.'

'You're right,' Fraser agreed. 'Because science, it does need explaining. Like, those diagrams you emailed—they did make about as much sense to me as...I don't know...*knitting* patterns.' He wondered if perhaps he should slow down on the drinking, worried he was coming across as rude.

But Anatol laughed rather than appearing mortally offended. 'I work with RNA and what I am trying to do is to find a way of getting a protein to behave a certain way. So for one of the projects I'm working on I use 'aptamers'. These are RNA sequences - the G, A, C and U letters in those diagrams I sent—and I am trying to get them to fold in and bind with the 'bad' proteins in the body and change the way they behave. They can also target amino acids, bacteria, viruses, whole cells. For another project, I use micro RNA, miRNAs. These match up with another RNA sequence and they can cleave it, cut the RNA so it doesn't produce a particular protein.'

'So you're sort of like Dr Frankenstein then?' Fraser teased.

'A *lovely* Dr Frankenstein! Using RNA, it is cheaper, more natural if you like than drugs. And the second project I mentioned, designing tuneable gene therapy systems, it will help with a neurological brain condition. The brain, it is very difficult to get to and RNA should make that easier. It will make people better.'

Fraser felt faintly ashamed of thinking of Anatol as a geek. There he was, intent on curing people of a debilitating illness, whereas Fraser had spent the last month writing about the ten best places to buy online cupcakes or devising quizzes to find out your celebrity spirit animal. Tongue sufficiently loosened by alcohol he confessed his hesitancy at taking on the assignment. Anatol gave him a playful tap on the arm and as they chatted Fraser realised just how much pleasure he was getting from the act of face-to-face conversation itself, appreciating what a fantastic a mode of communication humans had developed over thousands of years. Instead of stuttered glitches—'You're on mute! Gordon! You're on mute!'—someone suddenly disconnecting when they tried to pause the call for a toilet break, talking over others, it was a joy to pick up nuances, take a line of questioning because of the way an eyebrow had been raised, to follow the speed of thought. That and Anatol's hand gestures, the way he would pinch his lips when considering a particular point, rub his earlobe, the twitchiness of the man, all this he drank in. He wondered if the scientist's skills would be analytical enough to pick up that the flush to his cheeks wasn't caused entirely by alcohol. The rainbow Pride badge on Anatol's leather jacket hinted that they had more in common than just living in Edinburgh.

Fraser felt pleasantly giddy when they left the bar, Anatol explaining to him the biochemistry that was taking place to make him so light-headed. He picked up the sense of release in the air as they walked down Lothian Road towards Princes Street. Here was a city taking pleasure in being a city again, a space where diversity and intermingling and sharing cultures, touch and tongues, were encouraged, enjoyed with a near hysterical enthusiasm.

This year the parade to mark the beginning of the Festival season drew inspiration from the historical pageants Patrick Geddes, biologist, sociologist, urban planner and creator of the Outlook Tower, had tried to instigate although only one had taken place during his lifetime. Today's would celebrate the famous scientists Edinburgh had produced alongside unsung figures who'd helped advance scientific research or whose achievements should be more widely recognised. There were huge, animated walking puppets, mobile recreations of the statues fixed about Edinburgh in honour of such notable scientists as J.Y. Simpson and James Clerk Maxwell. Striding alongside them were the Edinburgh Seven, the first women undergraduates at any British university, as well as Srinivasa Ramanajum, the ingenious, innovative mathematician, first Indian Fellow voted into Trinity College Cambridge, and Mae Jemison, the first African-American astronaut. Around their feet sang a choir of cancer survivors praising Henrietta Lacks, whose cells had been taken and used for medical research without her consent.

'Oh, I read about her,' Fraser exclaimed, feeling chuffed at being able to show off his pop science knowledge.

'She is the source of the HeLa cell line. She is like, *immortal*! I use her cell line in my lab,' Anatol told Fraser who found himself impressed, as if Anatol had confided he worked with a celebrity.

'And this I like especially! Because it is a celebration of failure!'

Anatol pointed at the back of a large truck filled with people holding home-made sheep toys, two hundred and twenty seven of them, a reminder of all the attempts made before Dolly, the cloned sheep, was successfully produced. They clapped and cheered and followed the samba dancing NHS and social care workers, supermarket employees, delivery truck drivers, all celebrating the announcement that the minimum wage had been increased, recognising the contribution made by those who hadn't had the option of working from home.

'Come on,' Anatol took Fraser's arm and pulled him towards the Mound. 'We will go to Holyrood Park but take a detour first. I want to show you one of my favourite things in Edinburgh. Then you will see why I love science so much.'

Trying to keep pace with Anatol's long stride up the Mound, Fraser was reminded of the times he'd walked up there when the streets had been deserted. He'd been struck then by how much he'd noticed the details that otherwise he'd have missed: the faded signs on walls, seeing buildings as architecture rather than the services they provided, the closes leading off the Royal Mile that would otherwise have been obscured by visitors. The city had been laid bare and it had felt dream-like pacing through it, powered along by the sense he couldn't linger, turning himself ghostly as if it was his alone to enjoy. He felt a little guilty now for missing that as they battled their way through the visitors and tourists across the Royal Mile and along towards Chambers Street. It was a delight to enter the National Museum again, the soaring white metal and glass taking his breath away. He paused for a few seconds, absorbing and recognising the feeling of being fully aware of his surroundings, somewhere he had previously taken for granted, horribly conscious now of how easily things could be taken away.

'I wanted to show you this because of something you said back in the bar,' Anatol told him as he took him into the science section of the museum, 'about how you had a friend who would rather just have the magic of things, that they felt science 'reduced' everything down to facts. I want to show you some *magic*.'

They passed rockets, lighthouses, satellites, aerial transmitters, advances in artificial limbs, stopping finally to stare down at an unremarkable rectangular glass box. At first it seemed to contain only blackness and their reflections: Anatol's expression fascinated, and Fraser watching his delight, puzzled.

'This, this is my favourite thing. When I heard that this museum had a cloud chamber, it was one of the first places I visited.'

'What, is it something to do with meteorology or—' Fraser stopped when Anatol raised a finger, whispered, 'Watch,' and tapped the box's lid.

Just as Fraser was about to ask, 'What am I looking at?' he saw it. A small streak of what appeared to be grey smoke flashing through the black. Then another thinner, slower contrail, ski-tracks in the dark.

'Cosmic rays,' Anatol explained, his smile charmed by Fraser's bewilderment. 'They come from stars exploded, deep out in space, travelling all this way, travelling *through* us. Particles passing through you all the time and here they leave this. Like a breath. It is beautiful, isn't it? Lighting up this box...this is the wonderful thing about science,' Anatol near-whispered. 'Making the invisible, visible. Showing you incredible things in a box of mist.'

Fraser felt bits of his brain expanding, creaking into life, a range of new subject matter opening up to him. 'Imagine...' he murmured. 'They could have one of those things for emotions or the ways you're affecting people and you don't even notice it. You'd be standing next to it and see these sparks of red if you were feeling angry or annoyed or pink maybe if you were feeling...excited.'

Anatol smiled and Fraser imagined something invisible shooting through them, flowing through blood, bone and sinew, sparking the same reaction, flaring chemicals and electricity, leaving traces.

Outside the museum, Fraser rolled his eyes at the sight of a silent disco heading along to Victoria Street, following the lead of a guy with a bright lemon sweat-band wearing a day-glo orange t-shirt and green shorts, but Anatol tutted at his disdain.

'Watch them, I like the way they move. Copying the leader and each other, a bit behind the beat. All these separate parts making one thing, the dance, and they sort of become like an organism.

You know, when you come to my laboratory I will have to show you the cells I grow. To see we make them glow as green as that man's shorts! Using colouring from jellyfish—but that's a whole other thing. And when you look at them, they are so beautiful, like constellations, like looking at the stars! And this is amazing, but heart cells, they are like those dancers. They beat, they have a pulse. If you separate them out, they start to lose time, to get out of sync. Bring them together again and they synchronise again. Like they are dancing.'

'Honestly, if you're able to make me appreciate silent discos then yes, truly I will believe science can perform miracles. And I suppose,' Fraser contemplated the dancers disappearing down past Greyfriars Bobby, 'it is sort of what it was like a few months ago. Everyone separate but together, keeping in their own wee bubble.'

As they walked along to Holyrood Park, Anatol told Fraser how happy he'd been when his research had brought him to Edinburgh from Poland, how the size of the city was perfect, making it easy to make friends. He'd been looking forward to enjoying the Festivals the year before so when he'd had the opportunity to contribute to the events this year, he'd been thrilled. 'Such a good way of bringing people together, which is what cities should always do. Like you and me, by chance! Also, I think, scientists—we need to have more *time*. When we were in lockdown, yes, it was awful not being able to get to the lab but it was also a good thing too, in a way. I could think and read and plan all the experiments, the testing, the research, I would *like* to do rather than *have* to do. They might lead nowhere but that's alright. Because, see, failure is just as important, as useful as success in science!'

The Park next to Holyrood Palace was covered with marquees of varying sizes displaying the results of collaborations between artists, craft-workers, performers and scientists. The largest exhibition space was filled with the innovations that had taken place during the Covid-19 outbreak: the designer facemasks, the new form of hand-dance developed from handwashing techniques, the lightweight goggles made out of recycled yoghurt cartons, ventilators made by wind-farm technicians. Next to a large moving sculpture of plant forms created out of all the plastic used by labs, the ends of thousands of pipettes, melted down to become twisting vines, waving leaves, thin, floating petals, was another tent displaying knitwear inspired by the coding of DNA.

'This is a lot like the work I do,' Anatol explained. 'Like a sort of craft. Knitting and splicing. Stitching things together. And the computer programmes I use, with colours for each of the sequences of proteins, they create patterns, like these.' He pointed to his jumper lined with lilacs, pale greens, yellows. 'I am wearing myself! These patterns, they are from my DNA, my sequence.'

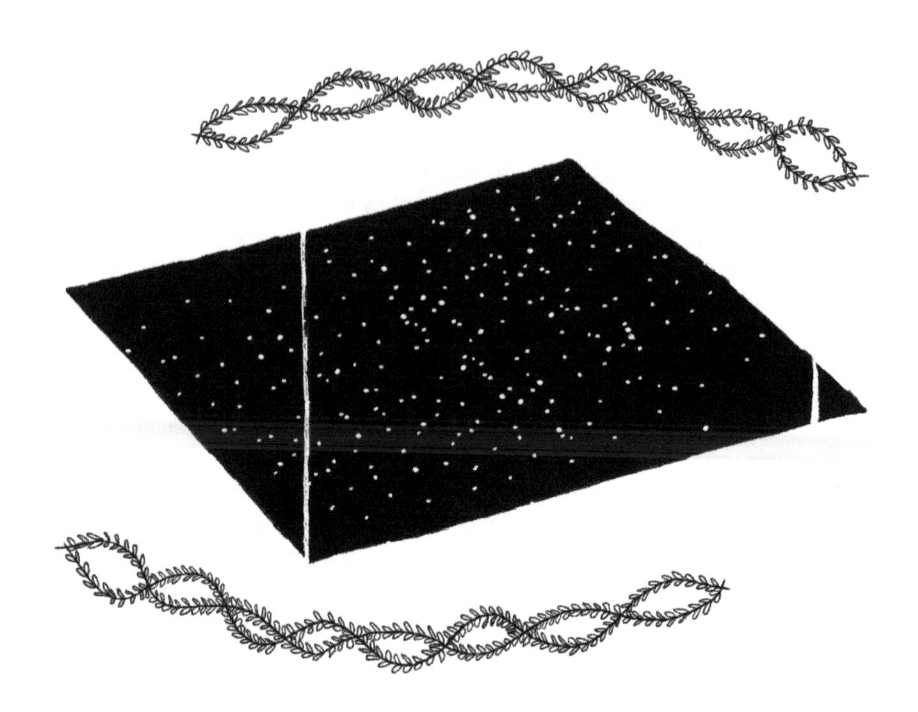

Fraser was suddenly reminded of all the jumpers his mum had knitted for him that had always managed to be that bit too tight, that bit too scratchy. Yet he'd worn them at home (they were taken off for the t-shirt underneath as soon as he headed outside) because of the look on her face once she'd tugged it down over his head: happiness revealed. When he and his sister had cleared her flat out after she'd died, he'd found them all carefully folded in a bin-bag at the bag of a cupboard. His and her DNA sewn into the wool, stitched together with love.

'Wearing your heart on your sleeve then,' he remarked.

Anatol gave a light clap of delight. 'That's it exactly! And here, now I must really show off to you. It's a bit of a walk, but you're up to it, yes?'

He led Fraser past St Mary's Loch, logging into a Festival app and holding up their smart phones so they were able to look back in time to see the volcano that created Arthur's Seat and the Crags erupting. Fraser could practically feel the heat from the lava scorching his brow, felt his step turn unsteady, imagining tectonic plates shifting beneath his feet. The landscape of his city turning strange again, unfamiliar, unpredictable, and he lowered his phone, felt a spike of panic at the thought of how quickly normality could vanish.

'See, all this science can make you unstable!' Anatol laughed, and leant against Fraser who disappointed himself by flinching away, the close proximity shocking after months of keeping a distance. Thankfully Anatol didn't seem to notice, started racing ahead until he could point excitedly to the markings decorating the side of Arthur's Seat by Dunsappie Loch: not virtual this time but actual huge chalk drawings, hundreds of feet tall in blue, red and green.

'Here, look, this is me! My work! You see, the sort of green stitching there, that's trying to get the RNA to do what we want it to do.'

They looked like the creation of early settlers, one resembling a crudely drawn man monster with splayed feet and thin tentacles for arms. The one facing it was the same form with its head pulled down by a band linking it to its left foot, one tentacle linked to the left thigh, the right foot twisted to face the other way, bound to the opposite heel.

Fraser knew he was anthropomorphising, attributing human characteristics to diagrams of genetic sequences but he couldn't help but be oddly moved by what he saw. The very building blocks of life looking like two figures caught in a strange dance or one seeing its reflection transformed, twisted into a different shape, reformed to perform a better function, and bound so it couldn't do damage. This was making microscopic healing visible and epic, showing how far technology had advanced over thousands of years; science, improving the way we lived, making things better.

'It's weird but...beautiful,' Fraser murmured but to himself as Anatol was already off, bouncing about on the springy, coiled variation of a bouncy castle that had been created to match the form of the crystal structure of a TetR-aptamer complex.

'A *what?*' Fraser shouted over at Anatol as he bounced from one spring to another.

'I will explain it to you another time, when we go to the lab perhaps!' Anatol laughed breathlessly back and Fraser smiled, not just because he was making an idiot of himself but because he had the promise of a lab visit to look forward to.

They decided not to take the miniature steam train celebrating James Watt back into town but walked instead towards Holyrood Palace, pausing to stick their heads inside the giant kaleidoscopes built to honour David Brewster and the climbing frame dotted with kids based on the Gray Graph invented by Marion Gray. At the end of Calton Terrace, they hesitated, the sound of the samba band from earlier drifting up towards them, the performers obviously having decided they wanted to continue the party for as long as they could. Just as Fraser wanted to draw this interview out.

'If you like, if you have time, I could take you to one of my favourite places,' he asked shyly and when Anatol replied, 'I would like that very much,' he felt happier than he had done in quite a while.

He tried not to wheeze too much, realising how unfit he was after weeks of staying in, tramping up the steep path to the top of Calton Hill, the view making the pain in his chest worthwhile. Once he got his breath back and they saw how far they'd walked, admired the view of the dead volcano, the streets filled with life again, people sitting on the grass, the benches, taking their time, drinking it in, Fraser said, 'I like how it's not that big a climb but you get to see the whole city. That and there's an art gallery here and an observatory too. Like our day together. Art and science.' He hoped he hadn't sounded too cheesy and then realised he didn't care. That the joy of the day should allow a little cheesiness.

'The way it should be,' Anatol agreed. 'Remember what I was saying earlier about cells? Human cells, if they are separated out, they grow more slowly. It's like they're pining. It's written in our genes how sociable we are, that we gain strength from each other. So I knew we'd get through the past few months.'

'Cleaving,' Fraser nodded. 'You mentioned it earlier in the bar. It's an odd word. Means a separation but also a connection. An auto-antonym it's called or a Janus word. Like the God with the two heads. Sorry,' Fraser laughed, apologetic at Anatol's confused expression.

'My English, still sometimes it needs some work!'

'No, no, it's me, being a word geek,' Fraser reassured him. 'So you can cleave to someone, though that makes you a bit clingy, or cleave like cut them in half. But it's sort of what happened over the past few months. We were all apart but trying to group up. Realising how much we missed one another.'

'Exactly! You see what we can both teach each other? Like cells, you put them in proximity and they stretch out as if they are trying to communicate.'

They stood and looked out at the city and Fraser remembered a walk he'd taken in the midst of the lockdown. He'd negotiated the awkward dances with fellow pedestrians, Edinburgh's wide pavements not quite wide enough. He hoped his guilt wouldn't show and cause him to be quizzed about his earlier trip to the supermarket for three bottles of wine. He hoped if challenged he would blurt the real reason why his walk felt essential, that he needed to reconnect with the city he loved, the city that had shaped him body and soul, hardened his calf muscles with its hilly streets, soaked into his lungs through the reek of the breweries; the city he tasted in its chippy sauce, heard in the bagpipes playing when he walked up the slope from Waverley Station or, if in August and heading out the back way up the escalators past the Mall, in the Peruvian pipe band.

Back then he'd looked down at the empty streets, the buses carrying only their driver, and the view had blurred, the pain in his throat not from the climb, thinking of the missing, those disappeared by the virus. With Edinburgh scoured like this, not being in contact with friends and family directly, only via the slightly averted gaze of a computer screen, it hit him how much he missed the *thereness* of them. He pined for his city and he realised now how much it was made of the people who filled it.

Standing next to Anatol it felt the most natural thing in the world to reach and take his hand. To feel the sweat of him, to draw into the heat of a stranger, cleaving to fit someone else's pattern. The kiss that followed stitching their life into the city that breathed and thrummed about them and held them close, knitted them together.

A Note on the Science

by Adam Mol and Vivek Senthivel

Kirsti's story is a witty self-referential portrayal of her interaction with us through our Biopolis collaboration. The story is unique in the sense that it is not based on science (like how *Interstellar* is based on theory of relativity, black holes, wormholes or how *Jurassic Park* is based on PCR and palaeontology), but rather based on the lifestyle of a scientist, drawing metaphors from the scientist's work and the impact of latest scientific fields on society.

The character portraying a scientist in the story is inspired by Adam, and highlights the perspective of a researcher working on a modern topic like synthetic biology, his take on city life, attitude towards meeting people from non-scientific backgrounds, enthusiasm towards his own work, etc.

Kirsti draws parallels between human relationships and biomolecular interactions, simultaneously depicting human behaviour in the language of RNA-protein interaction and anthropomorphising aptamer-ligand interaction. Parallels are traced between the synchronisation of beating heart cells and Silent Disco, the colour of shorts and GFP protein expressed in cells, and the sequence of RNA and a jumper pattern!

The impact of synthetic biology on society in a post-pandemic situation is highlighted at multiple levels, from the problem of managing lab waste, to government-based initiatives for new measures to prevent future pandemics, and, ultimately, to peoples' acceptance of the importance of science to society.

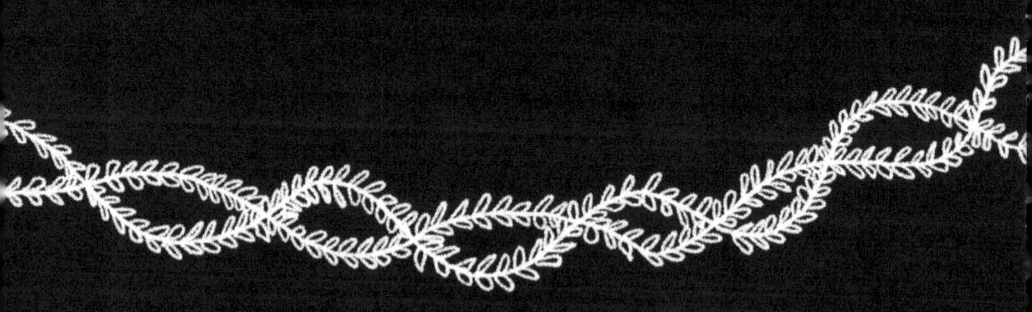

Kirsti Wishart has a PhD in Scottish Literature from the University of St Andrews and received a New Writers Bursary from the Scottish Arts Council in 2005. Her short stories have appeared in *New Writing Scotland*, *The Eildon Tree*, *404 Ink*, *Glasgow Review of Books*, *Product Magazine* and *The Seven Wonders of Scotland* anthology. She's been a Hawthornden Fellow, a contestant in Literary Death Match and is a regular contributor to *The One O'clock Gun*, a literary free-sheet once found in the darker recesses of Edinburgh pubs. Her first novel, *The Pocketbook Guide to Scottish Superheroes* will be published by Stirling Publishing.

Dr Adam Mol is a TRAIN@Ed Marie Skłodowska-Curie COFUND postdoctoral fellow working with Prof. Susan Rosser group at Mammalian Synthetic Biology Research Centre, University of Edinburgh. He completed his doctoral studies at Synthetic Genetic Circuits group headed by Prof. Beatrix Suess in Technical University of Darmstadt in Germany, as part of the Marie Skłodowska-Curie ITN MetaRNA network spanning the interface between RNA biology and metabolism at the single cell level. Adam is developing and applying the tools and methods of mammalian synthetic biology to unmet needs in basic biology and medicine. To build more robust genetic systems, Adam makes an effort to incorporate greater complexity into the design of synthetic circuits, such as developing more effective feedback control for therapeutic applications.

Dr Vivek Senthivel is a postdoctoral research associate at the School of Biological Sciences at the University of Edinburgh. His current research work involves developing receptor based biosensors for theranostics and drug testing in mammalian cell-lines. His background is in microbiology and most of his past work involved developing synthetic biology tools for bacterial and mammalian systems. Vivek's research interests involve pattern formation in biology, and topics like emergence, complex systems, network science and meta-analysis of scientific methods. He is keen to participate in public outreach of science and discussions around the influence of science on society.

Clare Duffy

The Second Brain Chain

Hello. My name is Dame Alison Barnard. *Dame* as of today. The press have been calling me 'Dame Do-Do', but I don't mind. I really don't. I'm used to the ignorance of the masses, the wilful ignorance, I often think.

It's just so much more fun to not care about the truth, isn't it? It's just so much easier not to bother considering quantifiable facts particularly significant. 'We're all going to die'—that's what they think. (That's one fact everyone does accept—death.) 'We should have as much fun as possible while we're alive, right?'—I understand that point of view. You know what, I actually think it's a defensible one. We *should* have as much fun as we can, while we're here. It's an amazing shot—life. An amazing shot. I appreciate it more than ever now, here at the end of my life. So what if people think my life's work was about shit? That really is much more fun than getting your head around the idea that I engineered bacteria to make molecules that make you a different, happier, better person. I don't care. I really don't. I'm Dame Second Brain Chain, with an outlet in every major city around the world and a private client list that includes *everyone*. I have nothing left to prove and I'm here just to say, 'Thank You', before the end.

Actually, maybe death isn't actually that certain anymore. But I have to confess I can't help laughing when people start explaining their plans for uploading their minds. It's so fashionable at the moment, all this Whole Brain Emulation. I know how hypocritical I'm being. People used to laugh at me, when I told them that changing the bacteria in their gut could cure their depression. My problem with the idea of an uploaded mind is that it wouldn't be connected to a gut. The gut is the second brain, but as my company's branding has it, '*SECOND, NOT SECONDARY*'. We're not able to think with the 'first brain' without the 'second brain'. So, if you upload your mind only from the first brain, what will it know? Not much in my opinion.

But that's not what makes me really laugh about these nut jobs. Why do they want to live forever? You don't want to read the same story forever, do you? What will be next? There may be another existence, one without consciousness, very probably. What would it be like to be earth? What would it be like to become a microbe? The unique contribution my work and my life has made to the 21st century has been at the intersection of the spiritual and the bacterial. Where do all human needs meet? In the gut. In the thinking, feeling belly.

Death is very close to me now. She is like my cat out there in the garden, which overlooks Portobello beach. She meanders through the spring flowers I planted with my son so many years ago. She comes in and sits in my lap and purrs a few times a day, just for comfort. Death and I are getting to know each other. When I'm history, my work will be sieved, spun at thousands of times the force of gravity and then all my ideas, all my life's work, my big turd of a contribution to science, if you like, will be reduced to a long list of individual bugs, and they will go off on their separate journeys, without me holding them together as one authorised story. But that's alright. That's what death is. It's the time to let go of the power you were granted for a century and some change. It's time to diversify.

So I have a few 'Thank You's I would like to make. My first 'Thank You' is to His Highness, the King. It was a pleasure to meet you again today, Sir. My medal materialised quite perfectly by my side and as you can see, I'm wearing it with pride. Of course, I am still anti-monarchist in principle. But I like status and recognition as much, if not more than, the next person. I'm at peace with my hypocrisy and at least it's an honest hypocrisy. I really want to say 'Thank You' for your public acknowledgment of how my work re-united you with your wife. You explained how our analysis of both of your biomes and our prescription of microbes restored your relationship. With a wicked glint in your eye, you urged the public not to 'poo-poo' my work and, in one fell swoop, it was the sensation of the century!

It was, of course, her, not you, who brought you both to my door. You were not convinced of the potential of the procedure at first. 'I loved her so much once, I would love to love her again,' you told me.

I told you to think of it as the most intimate connection you can make with your partner. Remember who you were when you met and what you did. Didn't you spend weeks entwined in each other's bodies? Isn't the process of falling in love first and foremost one of sharing microbes? On a first or second date you might brush each other's hands, kiss with pursed but closed lips, and in those first touches, exchange your skins' unique microbiomes. Your bacteria finding new places to live in the new person, changing them from who they were to who they would be. Those microbes sense they need to reach quorum and that propels you on to meet the person again, to touch more of those intimate areas that hold much more complex and diverse communities of creatures. These beautiful, microscopic, one-cell creatures are the carriers of huge need in humans. They make such huge decisions in our lives, and when there are enough of them, love blooms.

And in your case when love returned for you and your wife, your two heirs were the next step. The future of our country has been shaped by the creatures in your belly, much more than by your conscious decision to get married and have children. You loved your wife; I know that. Because when things cooled off, when recrimination replaced passion, when you were desperate and alone you came to me, just a humble professor in microbiology as I was then, working in my own time on the potential of spiritual and emotional healing through faecal transplants.

Of course, I didn't know it was you at first. It was your aide who delicately explained the situation and asked if I could help. Back then I ran a kind of 'back-street' clinic for the lonely, the fearful, the lost and excluded. I had created my first, rudimentary directory of the characteristic effects of different microbes on human emotions. I was first inspired as a young post-doc when the Farncombe Family Digestive Health Institute published their staggering and weirdly wonderful work in 2011. Those researchers proved that faecal transplants from bold adventurous Swiss mice could make fearful BALC/c mice 'go-getters', even 'jet-setters'! Immediately I had a vision, I foresaw the world as it is now, where you can easily analyse your unique microbiome and programme it to make you more generous, more calm, more loving, perhaps more like someone else or more like the person you once were.

But I guess we'll never know how many ways my tech can be used. I didn't think when I was returning the King and Queen to

conjugal bliss that there might be less noble reasons to make people fall in love with each other or to become different kinds of people. I've spoken out against the use of my technology to enslave so many. I'm disgusted by the idea that so many (mainly) women are microbially enslaved to (mostly) men and believe themselves to be happy. It was bad enough when I was a young woman, but at least I knew when I was being mistreated or abused.

(What can I do? I could have done so much more to keep the codes secure. But everything is hacked in the end. We are working so hard to regulate and criminalise un-lawful bacterial programming. But legislation, as we know, is always going to lag behind practice.)

But back to my 'Thank You', kind King. Thanks to you and your brave self-outing of the treatment you and your wife underwent, my twilight research became mainstream. We won prizes, we landed grants, I set up a department for the understanding of emotional and spiritual effects of bacteria and we started to programme the behavioural effects of bacteria in humans.

My second big 'Thank You', is owed to my business partner who shared my vision for a 'Second Brain' outlet in every city in the world. She came up with the name, the brand, our logo. The story of everything I have done. She is everything, the true 'second brain' of our company. Sandra, you have redefined more than anyone what we now understand 'thinking' to be. You have re-claimed the relationship between the first and second brains. Sandra, you are my number two.

(Sorry. I just couldn't help myself.)

But seriously, Sandra, you've been my inspiration for 40 years. When we first met, shortly after the King and Queen gave my discovery the royal thumbs up, you were so full of passion and ambition. You were intelligent, creative and hard-working. You had everything needed to make a success of anything you chose to do in life—except confidence. Our first project, the 'Yes You Can', was so successful because you really understood the need so many people have to let go of their anxiety and their fear of failure in order to succeed. Just imagine, you said, what the world would look like if the meek really did inherit it? What would happen if the meek actually believed they also deserved to be successful?

Microbes can't make you successful. Of course they can't. But they can allow your perspective on failure to change. Once you can experience failure as simply one potential outcome of taking a risk then it becomes a much less important consideration when making a decision. It doesn't have to be feared as evidence of your personal lack of worth. Our language shows we have always known fear of failure is generated deep in our gut. In minor cases it is known as 'having butterflies', in moderate cases we experience the 'colly-wobbles', and in severe cases it can be 'gut-wrenching', when we experience loose stools or vomiting. Irritable bowel syndrome can be a terrible, debilitating lifelong condition. The 'Yes You Can' meant the meek really did indeed inherit the earth. What a difference it has made!

Our second project went further by aiming to alleviate shame. This was less of an untrammelled success. Do you remember, my darling? In the early stages of development, how delighted we were with the improvement in the quality of life of victims of abuse? But it would have been a terrible thing to unleash on the general public. We had scaled up testing to only a hundred test subjects when we realised that this was a treatment that should never reach the mass market. We were both shocked at how little shame so many people feel about their wrongdoings. Wasn't it extraordinary to see how many people were in such denial about their faults, misdemeanours and crimes? The truism that 'no-one thinks they are the bad guy' proved to be so much truer than we had ever imagined. Our work proved how deeply complex an agent like shame is in human life. Shame is a bigger riddle than we were able to solve and so we have left that for the next generation to unknot.

I come now to a confession, my sweetheart. I have benefited from this crime for 20 years. I don't think what I have done has caused any harm—but in my gut I know it's wrong.

I cloned your unique microbiome.

When you decided to leave me 20 years ago now, I accepted your decision. I didn't do what those thugs, those dust mites of humanity do when they force feelings of love on people. I accepted you were no longer in love with me. I let you go, in the hopes that you would come back one day, in the hopes that I could still be in your life. But I also took you with me. I made an unauthorised biome-print of you and I used it sometimes, not so often now—but in the early days, when I really missed you, when I felt that life wasn't worth living without you, it might have been rather often. It was the 'you' of our first days of falling in love, when we couldn't be apart for even a moment, when you missed me if I went to another room. Can you believe that we ever felt that way?

Can you forgive me? Is there anything to forgive? Have I taken some part of your soul? I know I have broken trust between us. I've stolen. That is clearly wrong. But I haven't broken any law I know of. Oh but it feels wrong in my gut. Can you forgive me?

Well, here comes E.coli, my cat. I don't know how old she is. Her kittenhood is certainly long gone, but she seems so agile and comfortable in her body that I don't think she can be much more than eight or nine. She arrived one morning pawing at the garden door. I gave her breakfast and she stayed. I do like her to curl up in my lap when I'm tired. I enjoy feeling the vibrations of her purr ripple through my body. I feel like I'm letting go of my own rhythm and attuning to hers. I look out of the window at the marigolds and poppies; I listen to the waves breaking out of sight. I am without any particular prescription at the moment. I eat well. I have a belly fully of diverse creatures, and I don't feel the need, particularly, to change anything, anymore. Other than, of course, this final transplant I have on my table. A death transplant, or to put it more simply, just a poisonous plant in my afternoon tea.

This is my time to stop. So I want to say a last couple of 'Thank You's. First to my dear dear son. Jerry, honestly, this is the best thing for me. I am, as you have often pointed out, a really selfish old bat and this is in keeping with my character and life's work. Try not to be too angry. Hypocritically (again), I want you to accept that we are who we are. I never chose to take anything that would stop me being selfish, because I like getting my own way. Do you see? You are (I'm so proud to see) so generous and selfless. Your work, your life with your family is wonderful. If there is one thing I wish for you, my darling, it is a bit more fun. Let yourself be selfish now and again. Do what you want, just because you want it. It's okay not to care all the time.

If there were anything you wanted, I would give it to you. I love you. But of course I only feel like that when you are not here. I know we get on each other's nerves, but I have only ever had love for you, even when I want to clout you over the head with a frying pan!

I have kept my final 'Thank You', for Mr Suse. You will most likely be the one to find my body and I know that that is never a pleasant experience. I also know, Mr Suse, that you will deal with this better than probably anyone I know. There is an envelope in the key drawer. Take that in place of the apology I will not be able to give you. Please find E.coli a good home. My solicitors will take care of everything else.

Goodbye my friends! My second brain is gasping for that tea.

A Note on the Science

by Leonardo Rios Solis

The story explores some very interesting new scientific evidence about the bigger role and effect microorganisms that live inside humans have on our behaviour. It is estimated we have around 10-30 trillion microbes that live inside the human body collectively called the microbiome. The majority live in the gut and intestines, where their main function elucidated by scientists was for food digestion, synthesis of vitamins, and helping in the fight against infections. Nevertheless, recent findings are showing the role of the microbiome could be far more wide-reaching, including the regulation of appetite, mood and emotions, resulting in the label of the microbiome as 'the second brain'.

How the microbiome sends signals to the brain is still debated. Recent studies involving dozens of mice or humans have developed several theories including: a) Microbes could directly produce serotonin, which is a neurotransmitter largely known to regulate appetite and mood; b) Microbes could affect immune cells, prompting them to produce small proteins called cytokines, which could then travel through the bloodstream to interact with the brain cells; c) Microbes could produce specific metabolites such as gut-microbe-derived fatty acids, which have been found to signal the cells from the digestive tract to increase serotonin production.

Although the microbiome science is still young (10-15 years), there is already clear scientific evidence associating the microbiome with many diseases. Nevertheless, there is still work to be done to draw clear cause-and-effect conclusions about the vast inventory of microorganisms from the microbiome and its effect on human mood and emotions. In the near future, according to predictions from several researchers, it might be possible to deliver a dose of healthy microbes in the form of prebiotics, probiotics or even faecal transplants in order to promote a healthier operation of the human body, including positive emotions and behaviour.

Clare Duffy is a playwright and director. She specializes in data driven performances through her company Civic Digits, and is currently working on *The Big Data Show*, an immersive live performance, which blends digital tricks and games on the audiences' own phones while telling the story of the first prosecuted hack in the UK. She co-founded Unlimited Theatre in 1997, which blends theatre with scientific research. Clare won a Pearson Award for her first full-length play, *Crossings*, and a Platform 18 award at the Arches for writing and directing *Money the Game Show*. She wrote *Arctic Oil* as part of her fellowship at the Institute for Advanced Studies in the Humanities, which was presented at The Traverse Theatre in October 2018. Clare also wrote the Cbeebies' Christmas Show 2013 - 2019, adapted *A Midsummer Nights' Dream* for Cbeebies in 2016, which won the Royal Television Society's award for Best Children's Programme, and abridged *The Tempest* for Cbeebies.

Dr Leonardo Rios was awarded a lectureship at the Engineering School of the University of Edinburgh in 2017, where he leads his research group at the Institute for Bioengineering and the Centre for Synthetic and Systems Biology. His research motivation focuses on promoting further understanding on the interphase of Biochemical Engineering and Synthetic Biology as well as to find novel ways to apply this knowledge to tackle the growing socio-economic inequalities. In more detail, Dr Rios's research group focuses on applying synthetic biology tools to engineer microbial cell factories to produce high-value products such as biofuels, biomaterials and pharmaceuticals. They strongly believe that boosting the production and maximising the potential of their microbial cell factories requires the parallel development of novel bioprocesses to cater to the new qualities of the engineered cells. Dr Rios is also interested in entrepreneurial activities, and has co-founded the startups Marizca LTD and Logykopt.

Vicki Jarrett

Bloom

BloGen
BIO-REFINERY

Abram settled himself in the front room, with the blinds drawn. Sunlight filtered through the design of leaves and vines, casting the room in a wavering greenish light. A brighter green burst from the screen of the laptop as yet another archived episode of *The People's Vote* started its opening sequence. Abram noted down date, time and filename. He still doubted he'd find anything helpful in these old files but it paid to be thorough. He knew the thought that he could've missed a clue would play on in his mind if he skipped any potential evidence but these marathon viewing-sessions were taxing his patience. Had people really believed they were making a difference? That anything they 'decided' had any bearing on what was going to happen anyway? The choices weren't even real. The whole thing was theatre: mind games masquerading as democracy.

On screen, the lights in the auditorium dimmed as the show's host stepped up to the stage and gazed out at a sea of faces, all rapt in his direction. He rubbed his hands together and smiled, white teeth dazzling. 'Oh boy, do we have a vote for you today!' he exclaimed, and the crowd cheered and whooped.

'Every minute of every day, every second, every moment, we choose one possibility and not the other. We select one future over another. Infinite possibility, but only one path.'

He had the voice of an evangelical preacher, throwing down the words with rhythm and flair as he strode back and forth on the stage, intensity and sincerity emanating from every syllable, every movement, each arched eyebrow and raised hand precisely choreographed. The crowd was nodding along with each beat of his speech, the same speech he gave at the start of every show.

'Every time we try something, every time we decide, possibilities spin out in all directions. Dizzying, isn't it?'

On the backdrop behind the host, the graphics pulsed and spun, green Mandelbrot fronds swirling infinities in a way specifically designed to induce the necessary trance state for the upcoming segments. A thousand heads bobbed and swayed under their hypnotic influence.

And they all bought into this, thought Abram, shaking his head. Just opened their minds and let the authorities waltz in with their induced storylines. The past really was another country.

The host reached both arms out to his audience, the gesture part generosity, part benediction. 'Today we're going to look at three possibilities. Three possible fates, each one as real as the next.'

'And by that you mean not real at all,' said Abram out loud, surprising himself with how annoyed all this was making him. The blatant manipulation, he felt embarrassed for humanity, for his own ancestors and their placid gullibility. All of this only a hundred years ago. Too close for comfort. It made him wonder what aspect of the present day that he and everyone else currently accepted as normal would be subject to incredulous pity from some distant point in the future. He reminded himself that, watching the reruns onscreen, he was immune from the true pull of the induced storylines, meaning he experienced them merely as short dramatizations, whereas the live audiences in their immersive trance experienced the thoughts and emotions as if they were their own. Perhaps he should be more understanding, more forgiving.

'Three strands of possibility taken from the multitude that spring from the decision that lies before us as citizens. The bio-refinery planning permission is on the table. As the chosen thousand summoned here today to vote, you shoulder the responsibility to weigh this matter. Your thousand voices will speak for all, as one. You've read the reports, looked at the figures, the projections from the experts.'

In amongst the spiralling green, splashes of colour, fragments of graphs and pie charts bloomed briefly then died back, reabsorbed into the whole. Strings of numbers threaded and forked through the screen imagery like dark veins, pulsing with data.

'But ultimately, you decide. You the people. You choose! Do we choose this path, or do we choose another? Please don't try to vote before the final segment concludes. Your vote will not be registered.'

The Mandelbrot patterns swirled, pulling the vision inwards then downwards at stomach-dropping speed until, with a final rush of green:

FADE IN:

INT. SARAH AND ZAK'S BATHROOM

Aerial view down into a small, cheerfully cluttered bathroom. A high shelf crammed with houseplants trailing foliage down the walls. Steam rises.

A young woman (SARAH) lies in the bath, heavily pregnant, her belly and breasts smooth domes rising up from the water. She winces and runs a hand over her bump.

SARAH

(laughing) Easy, little sprout.
Let me have some peace while I

can. You'll put a stop to that
soon enough.

She raises her knees and slides down into the water, submerging her face slowly. The water rises past her ears, over her eyes, covering her mouth and nose last. Under the water, her hair snakes out, as if it has a life of its own.

Sound of a door closing elsewhere in the apartment.

INT. HALLWAY

A tall stooped young man (ZAK) locks the main door and throws his keys onto a hall table, noticing opened mail lying there. CLOSE UP of bill headed 'Green Zone Tax–Final Warning'. Zak throws it back down and rubs a hand over his face.

INT. SARAH AND ZAK'S BATHROOM

Zak enters and sits on the lowered toilet seat. He is wearing green-streaked overalls and his hair is matted with green dust.

SARAH

(surfacing) Thought I heard the
door. You look tired.

ZAK

You look beautiful.

SARAH

(laughs and flicks water at Zak)
Do you want me to leave you the
water?

ZAK

Yeah (stands and groans,
stretching arms above head).

SARAH

Hard shift?

ZAK
Ocht, no worse than any other.

INT. ALGAE PROCESSING DEPT

Noise of machinery. Industrial space clouded with green dust. People in overalls toil at various tasks, forklifts at work. A worker hefts a canvas sack onto a pallet. It lands in a cloud of green dust.

EXT. BACK OF BIO-REFINERY BUILDING

STEVIE (heavy-set, middle-aged) on a break, smokes and swigs from a bottle of green liquid. He looks out to sea, across the seaweed fields where several small boats rock in the choppy waves, workers in waterproofs haul up ropes of seaweed and cut away heavy ribbons as others wrestle them into dripping bales.

Zak rounds the building and approaches.

STEVIE
Alright big man?

ZAK
Still making the gut-rot? That stuff will kill you, you know.

Stevie offers the bottle to Zak who takes a swig, grimaces, hands it back.

ZAK
You got them?

STEVIE
Would I let you down?

Stevie hands Zak a paper bag. Zak extracts overalls caked in green algae, with the BioGen logo, identical to those Stevie wears. He pulls them on, rubs green dust into his hair etc.

STEVIE
How long you going to keep this up? You should just be honest with her. She'll understand.

ZAK
I can't. Not with the baby due. She'd freak.

STEVIE
And you think she'll freak less when the baby's here?

Zak shrugs, avoiding eye contact.

STEVIE (CONT'D)
Believe me, you do not want to find yourselves back in the brownfields with a new-born in tow.

ZAK
I'll find something else.

STEVIE
Yeah, well, you better. How long you got left on your green cards?

ZAK
We've got time. A while. Well, they run out soon. Like, next week actually.

STEVIE
(shocked) Ah shit, man. That's your transport, your rent, food, everything. You'll be exiled! You know they don't make exceptions.

ZAK
Yeah, yeah. I know. (sarcastic) Thanks for reminding me though, mate.

STEVIE
Sorry. It's just… What you gonna do?

ZAK
I dunno. Really I don't.

Stevie hands Zak the bottle again. Zak takes another swig and hands it back.

STEVIE
I've gotta get back.

ZAK
Yeah, course.

STEVIE
Look, good luck mate yeah?

ZAK
Yeah, thanks.

Stevie hugs Zak briefly then goes back inside the building. Zak stands and looks out to sea for a while, then walks away.

INT. SARAH AND ZAK'S BATHROOM

Sarah stands and takes the towel from Zak.

SARAH
D'you see the bill in the hall?

ZAK
Yeah, I saw it.

SARAH
I thought you'd paid it.

ZAK
I did. Don't worry. They've messed up somehow. I'll deal with it.

SARAH

(adopting lighter tone) I've been thinking about names again. It's about time we settled on something.

ZAK

I guess.

SARAH

I like Ethan, and James, and (wraps towel around her hair) Noah. That's my favourite.

ZAK

Bit biblical, no?

SARAH

Yeah, I know, but, well, I just like it, we don't have to decide right now. But soon.

ZAK

(lowering into the bath water with a groan.) Yeah, soon, soon.

Sarah lets a little of the bath water out and tops up with more hot, adding a splash of bubble bath. It's green, the bottle stamped with the BioGen logo. Filmy green bubbles mound up under the tap.

The bottle suddenly drops with a splash into the water.

ZAK

Hey! What're you—

Sarah gasps and grips the side of the bath.

ZAK

What? You're not, it's not… (leaps out of bath) Oh shit! What do I do? What do I do?

SARAH

(through gritted teeth) Hos…pital.

EXT. KERBSIDE OUTSIDE FLAT

Zak holds his arm in the doorway of a sleek driverless transport pod to stop it moving off as Sarah settles inside. He sits next to her and takes her hand. The pod moves off with a slight jolt and a hum.

EXT. CITY STREETS

They glide through clean tree-lined streets. At times it's hard to tell they're in a city. Vertical gardens and living walls clad every building, gardens on every rooftop. A few citizens stroll around, relaxed. Exotic fruits hang and flowers bloom in abundance. The BioGen logo appears here and there on billboards and signs.

SARAH
(placid between contractions)
It's so beautiful in the city now. Sometimes I forget how lucky we are.

ZAK
Sarah, I–

SARAH
When you think about what it was like before they built the refinery and sealed the green zone.

The transport sweeps on to the coast road, houses on one side, bio-refinery visible offshore on the other.

ZAK
There's something I–

SARAH
Can we stop a minute? (louder) Transport stop.

The transport pod glides smoothly to a stop.

SARAH
Windows down.

The windows softly lower. Sarah inhales deeply.

SARAH (CONT'D)
Night flowering Jasmine. So gorgeous.

ZAK
You know they only grow it here to cover the stink from the refinery.

SARAH
I know. But it works doesn't it? And having a purpose doesn't make it any less wonderful.

ZAK
Look, there's something I need to tell–

SARAH
Shhh. Don't. Everything's perfect. We have a beautiful home, we're healthy, soon we'll be a family. We're happy. That's enough for now. Isn't it?

The sun goes down along the coastal road as the transport moves off again. The Bio-refinery glows in an emerald haze. Fields of seaweed stretch out behind it. Marker buoys catch the last of the sun's rays, studding the sea with flashes of gold all the way to the horizon.

Green and gold blurred and swirled blending the view of the coast road, the bio-refinery, the seaweed fields, along with Zak and Sarah's story into nothing more than patterned shapes and colours that swallowed their reality and their uncertain fates whole.

Abram rolled his eyes. The host was back, strutting and declaiming while the audience's gaze followed his progress from left to right across the stage with the dumb concentration of the concussed following a doctor's moving finger.

'No one is promising perfection. The future is a balancing act, as life always is. Nothing is without impact. There are always compromises. What sacrifices are we willing to make? What will we achieve and what will we tolerate? On what scales, by what measure do we judge the weight of our choices?'

The green and gold swirls on the screen began to slow, to take on deeper hues. The sea again? Abram hunched forward, trying to make out the details. Yes, definitely the sea, and was that the bio-refinery again too?

FADE IN:

EXT. THE COAST ROAD

Five human figures (WILSON, RHYS, ANDERSON, STEWART and KENNEDY) emerge from a green-gold dusty haze. They have scarves wrapped around their noses and mouths and wear aviator-type goggles, large backpacks and are armed.

Everything on the street is thickly coated in green algae. At a loud screech STEWART points his rifle upwards at a bulky green pigeon perched on a lamppost. Its eyes are red.

STEWART
(taking aim) Ugly son of a—

A shadow falls over STEWART and a far larger raptor-like bird, also with green plumage, swoops down and carries the pigeon off (mid-squawk) in its talons.

The company watch it fly off into the distance.

ANDERSON
Woah. That's the biggest so far.

KENNEDY
What even was that?

WILSON
Okay folks. Keep it together.
Let's go. (points)

Offshore, connected by a long thin pier, the domes and blocks of a dilapidated bio-refinery, streaked with green and black algae rise from the sea. Slow waves heave themselves up onto the shore with a rolling squelch, the sea a thick seaweed soup.

The company proceeds onto the pier, eyeing the water. Stewart abruptly pulls his scarf down and vomits off the side. Wilson pats him on the back.

STEWART

(spits) I thought the stink couldn't get any worse.

WILSON

Not far now, come on.

Stewart wipes his eyes and pulls his scarf firmly back over his mouth and nose. The sea squirms, something large moving through it. The company pick up speed and reach the bio-refinery.

Two stand guard while the others prise a board off the main doors and cut through a chain on the handles. Above the doors, the BioGen logo is visible.

INT. CONTROL ROOM

Inside they make their way to the control room. Kennedy unpacks a field radio.

WILSON

(into handset) Recovery team Delta reporting. We are on site.

RADIO V/O

Proceed with caution, Delta. Report in one hour.

WILSON

Will do. Delta out. (replaces handset) Rhys, Kennedy - secure the perimeter. Stewart, Anderson - let's get the backup generator online. We need power if we're going to access these systems.

Rhys and Kennedy depart, the others move around the control room, which is equipped with banks of displays and consoles. Wilson brushes dust off the main one.

EXT. OUTSIDE BIO-REFINERY BUILDING

Kennedy and Rhys split up and circle the building in opposite directions.

The sea heaves and bubbles. Kennedy swings around pointing his rifle wildly but no clear target presents itself as the sea settles again.

Rhys inspects an external door that has been forced. The metal is buckled, dark green mossy matter with streaks of red smeared around the open edge and on the ground leading inside. She runs a gloved finger through the matter and squints at it.

Kennedy, intent on some dark green splashing the walkway, stops to take a sample. He scrapes a small amount into a test tube and seals it in a clear zip lock bag. CLOSE UP the substance pulsates and shimmers.

Rhys, still staring at the substance on her gloved finger, brings it close to her face. As she watches, a slender pin of a shoot grows out of it with a tiny red seedpod on the tip, which bursts, releasing a fine spray of bright green-gold pollen.

RHYS

(sneezes) Ugh. Shit. What the fuck?

Rhys wipes her glove on her overalls and peers into the darkness through the doorway, takes out a walkie-talkie.

RHYS (CONT'D)

(into walkie-talkie) There's a doorway been forced here on the

north side. Looks like some kind
of storage unit attached to the
main building. I'm going in to
check it out.

KENNEDY (V/O)
Roger that. I'll circle around.

Rhys squeezes through the partially open door.

INT. CONTROL ROOM

Back-up power is on. Banks of dimly flashing lights and screens. Wilson stands next to Anderson seated at the main console. Onscreen is a magnified video of molecular activity.

WILSON
What am I looking at here?

ANDERSON
Looks like some kind of algal
pathogen. But there's something…

WILSON
Something what, Anderson?

ANDERSON
If these are Pelagic Sargassum
cells—

WILSON
In English please.

ANDERSON
(sigh) Seaweed. These are
seaweed and this is the
pathogen… Look, here, these
should be dead sar- seaweed
cells, right?

Anderson zooms the display into a small area of dark green cell matter. At first it seems static but then starts to move and pulse, streaks of red appearing as the cells multiply and spread, swamping both the remaining healthy seaweed cells and those of the pathogen until it fills the screen.

EXT. OUTSIDE BIO-REFINERY BUILDING

Kennedy arrives outside the buckled door, looks around then goes inside.

INT. BIO-REFINERY BUILDING (DARK INDETERMINATE INTERIOR)

Kennedy walks cautiously through shadowy storage room. Something moves suddenly ahead in the darkness. There is an impression of size and speed but no visuals.

KENNEDY

Rhys?

Kennedy proceeds further into the interior of the refinery as low red emergency lighting flickers on.

In a corner of the storage room, Rhys sits slumped against the wall. She appears lifeless at first, head bowed to chest. In CLOSE UP the veins on her neck squirm and swell.

INT. CONTROL ROOM

Wilson, Stewart, and Anderson are gathered around a table strewn with maps and papers.

ANDERSON

This is the Great Atlantic Sargassum Belt in the summer of 2020, extending from West Africa to the Gulf of Mexico. That's approximately 35 million metric tons of Sargassum biomass (shuffles through papers) and this is global Sargassum bloom taken from satellite imagery in 2022. By then we're up to 200 million. And of course with the addition of farmed biomass–

WILSON

Get to the point, Anderson.

ANDERSON

Okay, so I've charted the bloom patterns over the following twenty years, which takes us up to now.

Anderson moves to a keyboard, rattles at the keys and an animated world map appears onscreen with areas of algal bloom in bright green. The areas grow and shrink, grow and shrink, but always increasing in total mass and linking up.

WILSON

Hang on.

Wilson goes back to the earlier display with the molecular activity and runs the sequence again. The patterns synchronise with each other, moving to the same hypnotic pulsing rhythm.

The group's concentration is broken by a noise of movement from outside the control room. They exchange glances.

WILSON

(into walkie-talkie) Rhys? Kennedy? Come in. (static) Come in! (static)

INT. BIO-REFINERY BUILDING (DARK INDETERMINATE INTERIOR)

Rhys, her skin almost luminous in its paleness, sits with the body of Kennedy next to her, his head resting on her lap. She strokes his hair and hums a nursery rhyme. Her eyes are deep green. CLOSE UP her irises ripple and pulse with flecks of red.

A dark vein pulses in Kennedy's temple and his eyes move under his eyelids.

RHYS
Shh. There, everything's okay.
You'll understand soon.

Kennedy opens his eyes. They are the same shade of green as Rhys'. They look at each other lovingly. Kennedy raises a hand, Rhys presses her palm to his, dark algal strands twist around their joined hands as they both smile and sigh.

KENNEDY
It's like I'm… (searching for right word)

RHYS
Home?

KENNEDY
Yes.

RHYS
We're all going home now.
Finally.

KENNEDY
We must tell the others.

RHYS
(laughs, delighted) They're going to be so happy.

They clasp their other hands, stand and press their bodies close. Soon they are completely bound together in green strands that pulse and twist as the two bodies merge together.

INT. CONTROL ROOM

The radio set blinks into life in the deserted control room.

RADIO V/O
Come in Delta. Come in. Come in.

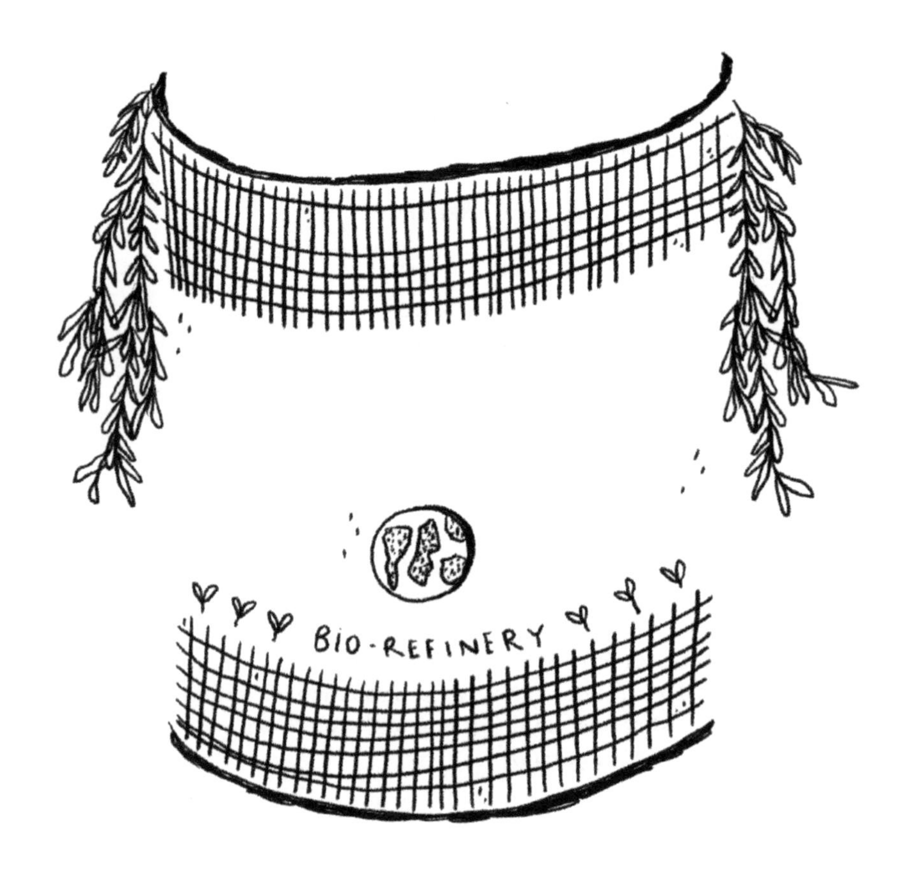
BIO-REFINERY

Abram snorted derisively. Good grief, they were really reaching with that one. More science fiction movie than possible future reality. Who wrote these things? Were they tripping or what?

The host, now back on screen, was galvanised. Light reflected from his brilliant teeth, his face shone as he flung more looping sentences out to the audience.

Give them enough rope, thought Abram.

'Things may not be as they seem. Utopia may harbour the seeds of darkness and the darkest dystopia may yet bring forth the brightest green shoots of hope. How can we say which is better? We reach, always, for understanding, for the best we can do at any given time, with whatever knowledge and tools we have available to us.'

Some of the audience were actually drooling now, Abram noticed with disgust; wet chins glistening under slack mouths.

'You, the people, bear this responsibility. It cannot be left to presidents, prime ministers, or governments. We must act together to shape our future. We are the authors of our own story. We shape the narrative. Ultimately, our stories are all we have. They are everything that we are. Past, present and future.'

Abram watched as the screen started its rotation again, colours blending, swirling, mutating.

FADE IN:

EXT: POLLUTED CITYSCAPE

Greys and blacks, nothing green in sight, thick smog, between grime-streaked buildings gridlocked traffic pumps out fumes, people wearing breathing apparatus scurry, heads down.

Zooming in on a block of flats towards a high window shaded by a blind with a design of leaves and vines. Through the blind, into the room. A man (ABRAM) sits hunched staring at a laptop screen.

ABRAM

What the fuck?

A Note on the Science
by Abdelrahman Saleh Zaky

Global warming has reached an alarming level of 1.3°C above pre-industrial levels and is increasing yearly. This is largely due to carbon emissions from fossil fuels that we rely on for energy, especially for transportation. If we do not act quickly and collectively, global warming could exceed an increase of 4°C by 2100. This will lead to catastrophic and irreversible climate change, including severe drought and rising sea levels, resulting in severe shortages of water and food, disappearance of cities, and extinction of many species of living organisms. Hence, global carbon emissions must rapidly decrease to net-zero by 2050, then further decrease to a negative value to stay within the safe limits (1.5°C) established by the Paris Agreement in 2016. This requires us to replace fossil fuels with clean energy sources. Among many alternatives, biofuels are an attractive option because the production process has a great capacity for carbon capture and storage. However, we do not have enough arable land and freshwater to grow enough biomass for biofuel production to satisfy the likely demand as well as capturing the carbon already released into the atmosphere. A UK-based scientist has proposed a 'coastal integrated marine biorefinery system' for the simultaneous production of biofuels and other valuable products with negative carbon and water footprints. The core of the project will be the cultivation of seaweed in huge marine farms covering hundreds of square miles of ocean. The system will not require the use of arable land and freshwater throughout the production chain, and will be linked to offshore renewable energy sources to increase its economic and environmental value. This will result in the capture of trillions of tons of atmospheric CO2 and the production of billions of tons of biofuels (liquid and gas) needed for transportation, as well as high value chemicals, food and animal feed products.

Vicki Jarrett is a novelist and short story writer from Edinburgh. Her first novel *Nothing is Heavy* was shortlisted for the Saltire Society Scottish First Book of the Year 2013. Her collection of short stories, *The Way Out*, published in 2015 was longlisted for the Frank O'Connor International Short Story Award, and shortlisted for the Jerwood Fiction Uncovered Prize and the Edge Hill Short Story Prize. Her latest novel, *Always North*, which moves into more speculative territory than her previous work, came out at the end of 2019 from Unsung Stories and was shortlisted for the Red Tentacle (Novel) Award in the Kitschies 2019.

Dr Abdelrahman Zaky is a research fellow at the School of Biological Sciences, University of Edinburgh. He is the Managing Editor of the journal Biomass and Bioenergy, and Chair of the Early Career Researchers network of Supergen Bioenergy Hub, funded by the UK BBSRC and EBSRC to work with academia, industry, government and societal stakeholders to develop sustainable bioenergy systems. Dr Zaky has published several papers on seawater-based biorefineries. His research aims to establish a Coastal-Marine-Biorefinery-System (CMBS) that relies on marine resources (seawater, marine biomass & marine microorganisms) for the efficient and simultaneous production of biofuels and High-Value-Chemicals. The proposed system will refrain from using arable land and freshwater throughout the production chain and potentially achieve negative water footprint and negative carbon emission values. It will be linked with other costal-based renewable energy sources (e.g. tide, offshore-wind, and solar) to improve the overall efficiency and economic feasibility of the system.

Viccy Adams

Aliyah Walks

NOTES

The Mayfield parish church marks the point where she breathes from the pit of her belly again. Shoulders lowering and back unhunching. The clog of rust and smoke in her throat fading. Some of the overhead halogens still work here. Aliyah passes from one pool of brightness to the next, exchanging nods with the other solitary women who stride by, humped by nylon rucksacks in a multitude of fading colours or tilted slightly to one side by swinging briefcases and patched-up messenger bags.

She eyes the tell-tale bulges of tins in one passing shoulder bag with instinctive envy, mouth wet with thick saliva as she pictures the woman—whose face is obscured by a large purple mohair shawl—gobbling stew so hot it burns the tongue and lights on fire all the way down the gut. Then Aliyah misses her footing in a shallow pothole in the tarmac and falls, arms springing forward to brace.

The notebook flies from her hand into the darkness of a wet hedgerow. Aliyah curses as she ducks her head to push through and scrabble for it, a branch scratching the length of her arm as her coat sleeve tears. The disturbed shrub smells intensely organic. Then her nose catches a dark, metallic scent and for a heartbeat she stops breathing, paralysed.

Although she knows it won't help, Aliyah opens her eyes. Through the leaves she sees the serrated outline of a building looming past, wires and plumbing shedding dirt and crumbled concrete. *Another one. A fresh one.* Something glows red up there where a person would have eyes. It's heading the wrong way to be joining the nest at the airport. *This is it*, she thinks. *They're coming into the city again. I'm going to have to move. The lab is going to close again. This will all have been for nothing. I've been too slow.*

Then she is momentarily blinded by the jolting beams of torches. A clump of people running, their outlines skewed by hard-hats, thick gloves, padded boiler suits. Their equipment slows them down, makes them stumble. Exultation rumbles through her veins that *they* are what the building is after. The people run out of the pool of light cast by the halogens from the pavement. Aliyah can see nothing of them now but those wild swinging beams, an echo of the blood pounding in her throat. *Maybe they'll make it somewhere safe.* A door in the lower part of the building swings open and something reaches out—barely visible, she's imagining it perhaps. But—yes, all the beams of light have disappeared. The crimson glow flares and the building begins its roar of satiation.

Aliyah tries to cover her ears but one hand is trapped by branches. She panics and tugs at it. Her arm frees. As she slides it back, her fingers catch hold of a wire spiral and then she's backing onto the pavement, clasping the notebook to her chest. One corner bent upwards and the whole thing damp but her precious data recovered. Aliyah pushes her hair back under her hood: the drizzle has intensified. The hairs on the back of her neck prickle and she sets herself towards the city and picks up the pace. She pays closer attention to the scuffed ground, chanting mantras of relief in her head with each uneven step.

By the time she's putting her key in the big old lock of her tenement, she can laugh at herself. She keeps one hand to the wall as she rushes up the half-lit stairwell, pictures the frantic spider she must have looked to any passersby as she clawed her way through the thin veil of dark leaves. The flimsy, plywood door of her apartment closes behind her and she rests her face against the cool grey paint of the bare studio walls.

Imagine that had been a few blocks further back. Where a bigger building might have still had an appetite. There wouldn't have been enough light to see where the notebook went. Or if I'd broken an ankle, running. If. If. If.

After her breathing has calmed, she puts a slice in the toaster, giddy and famished by the dropping adrenaline. There's a brown smear next to the scratch on her arm, black marks muddying down both sides of her coat as if she's been embraced by a puddle. Her mind brings up the images again and again and each time she tries to remember the details. *Georgian villa. Sandstone. Probably terraced originally, by the crumbling lines down the sides. A small, jutting Juliet balcony marring the symmetry of the façade. False balustrading detail at the top. Not that fancy, comparatively. Should I bother reporting it, or trust that someone else will?*

The evening news update confirms further signs that a third wave is cresting. Smaller properties, second homes. Aliyah switches to one of the pre-recorded channels and listens to old adverts for luxury goods while she mends the coat and writes up the day's observations in the recovered notebook.

Later, sitting in the shallow tin bath by the window, she runs a tentative hand over the experiment on her inner thigh. It has grown again since she checked it last, pressed up against the cubicle door at lunchtime and zoning out from the chatter of other women washing their hands. Aliyah wonders if it heard the building roar and responded—was there a twinge of recognition in her leg while she was trapped in the hedge?

Her fingers trace the grey dimples *there*, the sharp jut of red squares *here*. She wants to press in the hollows that she thinks of as windows but the deferred pain from the fall—the throbbing at every jolt of movement, the increasing heaviness weighing down one half of her body—has exhausted her. She dries off and writes down the measurements, takes her pulse again. Makes the final injection of the day and then falls asleep with still-damp hair, barely remembering to swap out the duvet for her grey towel.

The larger centrifuge by the main entrance is out of action. Again. Aliyah lingers over the curling sign on the table next to it, the handwriting and misspelling of someone she has never met in bold, angry capitals:

REMEBER TO SWITCH ME OFF AT THE WALL OR SUFER THE CONSEQUENCE

The flickering light in the long corridor that leads down to the inconveniently located, single toilet reserved for the women in the faculty. How ridiculous it is that none of them just use the closer, larger men's toilets, even now there are so few men left in the labs. She'll bring that up again at a team meeting, Aliyah thinks. Or maybe she'll just write a sign of her own and tape it up over the rusted steel cutout male outline:

TOILETS FOR EVERYONE

There is another note, on her workbench. *See me at ten.* Professor Eskbank's distinctive, looping handwriting. Aliyah is going to be late, no time to feed the cell cultures first. As she swings her bag to the floor she loses her balance and her side slams into the edge of the laminate. Her hands fly protectively to the experiment on her thigh—hidden by the long tunic she wears over her baggy trousers—and she has to force them to slap down on the floor instead. She pushes herself up, trying to act as if everything is normal despite the heat in her cheeks.

Updating her boss, Aliyah pulls out her stack of research notebooks and when she flicks through to find her meticulously labelled tables of the team's latest cell counts, it falls open to a page full of small twigs.

'Oh, it's nothing. I dropped it.' She leaves out the when and where, laughs it off as a funny anecdote. She stuffs the notebook back in her bag, finds the right one—growth figures for her official study.

'All data sets should stay in the lab.' Professor Esbank's—Julia's—mouth is tight and small.

Cat's arse, Aliyah thinks. She tries hard not to say it out loud, or to apologise. She's allowed to port the data. Everyone does it.

Julia's eyes scan the horizon then follow the curves of Aliyah's body like she's an insult. 'Lock it in a bag next time. Think of the risk, Aliyah. It's a bloody selfish thing to do to us, carrying it round on a jaunt for no reason.'

The taste of copper floods her mouth where she has bitten the inside of her cheek. *Eyes down. Don't react.* Her leg—the one with the experiment—judders and she clutches her bag more tightly on her lap.

With a brisk nod, Julia moves on to discuss the blood types they will need for the next stage of trials. 'I'll come with you on Friday to see the pair of doctors from the clinic. I'm due to touch base with them anyway; it'll save all of us a trip. What time are they scheduled for?'

'There's no need—really. They might not come.' Aliyah can feel the lie burn in her mouth. 'They said so last time. A student—they might send a medical student to take orders instead. Shortages again.'

Professor Eskbank's eyebrows arch towards her hairline. 'As if it's worth them running the usual surgeries. The sick are too useful these days.' She leans forward and smiles at Aliyah, unexpectedly. 'Cheaper than actual bait, right?'

Aliyah laughs a little too loudly. She has never felt comfortable with the casual siphoning of sick bodies from the clinics to appease the buildings, even back when it started and it was only the terminally ill.

Julia pats the documents on the table between them and a waft of her perfume scents the room with clove and spice. 'Fresh from the continent. These are the ones you asked for, yes? Sally looked them out for you, blame her if they're jumbled.' She frowns slightly. 'I do wish the doctors had a little more forbearance, though. We could do with the odd patient left for trials. Mention it, will you, on Friday.'

Aliyah nods and feels sick. Under her hand, her thigh is unusually hot. There is also a new tingling creeping from the edges. One of her first studies—under Julia's auspices—used donations from prisons and it makes her wince when she remembers how blasé she was about her colleagues questioning the volunteerism model. The dark shadow that now casts over all of them.

On her way out, Professor Eskbank turns a little and squints back from the doorway. 'Be more careful with the data, Aliyah. It's all we have now. We're all they have now.'

In the streaming sunlight, with a stomach still full of mid-morning porridge, Aliyah laughs it off again to her colleagues and explains what happened with the notebook as if it were nothing. They shuffle along the queue together, fingers picking over the small bags of dried pulses and dented cans. All the food is arrayed on the heavy wooden lab benches that get dragged off the old barricade for the monthly pop-up market, when the weather permits. Above them, the stone carved lettering of the ASHWORTH BUILDING has been picked clean of lichen again.

'Professor Eskbank is always like that.' Sandra's hair is pulled back tight, a blend of grey and black threads showing beneath her hat. 'Not surprising considering...'

The others shush her, covering their own mouths with their hands. Aliyah presses them but nobody is willing to talk. She tries to think nothing of it. When the armoured van turns down from Esslemont Road with fresh vegetables and a crowd forms and surges as if from nowhere, she finds herself grabbing the shoulder of the woman in front, jerking her backwards as if to expose her throat.

Something deeper than her stomach tells her to *bite*. Aliyah thinks she can hear the blood pounding through the veins of everyone pressing around her in the crowd. She struggles to remember what she is doing there. *Vegetables. I want vegetables.*

Sandra, arms bursting with a mixture of asparagus and kale, pulls her into the shade of a side entrance. She looks at Aliyah's queasily pale face and shaking hands and tuts, misinterpreting the reasons for them. 'Someone should have told you,' she says while her own hands sort through the bundle of greenery for slugs before she packs her bag. 'Julia's a bitch. She's risen fast in a vacuum, know what I mean?'

Back at her own workbench, Aliyah can't settle. Her hands continue to shake as she dusts off the glass bottles on the high shelf. She imagines sweeping them all onto the concrete floor, the brilliant noise it would make. How everyone would stare. She wonders if there would be any offers of help to clean up.

She was just an undergraduate when Professor Eskbank—Dr Eskbank, as Julia was then—recruited her for the T-cells Cancer study. Just the financial leg up Aliyah needed and what she thought was an unparalleled opportunity for hands-on experience. Then she discovered she'd been recruited because Julia could get away with paying her less than a proper post-doc. Most of her work turned out to be counting cell cultures with a hand-clicker. Typing up the scribbles of the more seasoned researchers who refused to use the assigned computers because they were too slow. Ignoring the jitters in her mind about *deserving better*, Aliyah began to read more deeply into the patterns she noticed emerging from their work.

Her classmates moved on. She worked slowly through her undergrad studies part-time, progress frustratingly slow while she tried to make rent with night shifts at take-away restaurants. She read everything on a tablet in those days, propped behind the counter between taking orders and handing over foil-wrapped packets of noodles. The ink of Julia's signature on her DPhil acceptance letter was barely dry when the data-enriched buildings Aliyah fantasised about living in one day evolved themselves beyond anticipating their inhabitants' needs. The world she knew began to disappear down concrete-lined gullets.

Her workbench then looked much as it does now—dust, containers, notebooks. Aliyah takes a cloth and makes a half-hearted effort to clean up: the cleaners do the bathrooms and the main corridors still, but they refuse to come in through the sealed doors. She stacks the scratched orange plastic culture bottles behind the methylated spirits in glass jars. She ought to be in the Clean Room, feeding those cell cultures. She can't let another batch die from apathy.

Julia was right. The sick are too useful. Aliyah cannot see the point in them continuing to look for cures when there is nobody left ill long enough to take part in trials, let alone to be cured. Sandra was right too—Julia would never have made Professor if it wasn't for the fact that her notorious underspending on equipment meant she wasn't as reliant on the technologies that were so swiftly banned. Even her research team—equally notoriously underpaid—were accidentally future-proofed by not being able to afford to live in the first areas to be connected to 7G. And if the rest of the Senior Faculty—better paid, better connected—hadn't taken advantage of the staff discount on the new smart technologies that could be injected into the walls of their homes to enable the fridge to know when you were hungry, the bed to make itself, the door to unlock when it sensed your footstep on the path. Aliyah has benefitted from Julia being a bitch—she still has a job. Promotion, even, a new wave of undergrads available to sit and click in the Clean Room while Aliyah reads journal articles and fantasises about turning back time.

Paths rolling up like tongues. Aliyah shivers and gives up the pretence that she is working. She passes back through the Clean Room and sees that someone else has fed her cell cultures for her—the new junior, she supposes. The new *her*. She punches her card to register that she is now absent from the building and joins the queue of other scientists waiting to leave early. Familiar faces, but not ones she normally talks to. The slots aren't set, but she's usually one of the last to leave—no school collection, no public transport loop to complete. The exhaustion today is overwhelming. She wants the privacy of her own home too, to peel back the fabric layers and see how the experiment looks.

How she wishes she could get on one of the occasional buses that clatter over the swept debris, with a rusting fender hanging off one edge and long claw-scrapes down the side. But none of the buses pass through The Dead Zone.

Aliyah smoothes her tunic down cautiously, trying not to let her fingers stretch towards the heat on her thigh, even though she can't see anyone looking at her. She sets her shoulders squarely and begins to walk down the remains of Mayfield Road, eyes straight ahead. The air smells of blossom and she can hear birds calling as they flock to roost. Her lip twitches as the clouds scud by, the orange streaks widening across the horizon. Aliyah remembers how much she dislikes The Dead Zone when it is light: the crowding of the rotting foundations where buildings have left, the sludgy green moss that seems to yawn and reach for you. It's just her imagination, but there is a constant flicker of movement wherever she isn't looking, as if shadows are scuttling behind the remaining twenty or thirty buildings that dot the half-broken, formerly desirable residential district.

In the crowd at the pop-up market today she thought she felt something in her thigh reach towards the people standing near her. She thought she could hear their blood thudding in their veins. *How contained is it? Are other people safe around me? Am I safe around myself?* Sweat is breaking out on her forehead. Her leg feels twice as heavy as this morning but, curiously, the pain is less now the heat has taken over. *Perhaps Friday is too soon. Maybe I should delay for another week, gather more data.* Aliyah wonders if she is more scared that the experiment will work than that it might fail.

Resisting the temptation to stop and check the spread across her thigh again, Aliyah focuses on the small stick of a woman walking ahead of her. Dark brown coat, thick boots, rucksack, dark hat covering their hair. She has no idea who it is—a friend, disguised out of lab clothes? Professor Eskbank herself? A visitor from another institution, off sightseeing in the city?—and wonders if she looks just as stiff and anonymous to the next person along. *It's almost as if we're expendable. The scientists, ready to be picked off, one by one. Just like the sick. The elderly. As if we deserve it.*

She matches her pace to the boots of the woman in front and distracts herself from the creaks and shifts around her by guessing what the woman bought in the market today. The dusk has fallen enough to mask the nature of the leafy greens peeking from the flap of the rucksack.

Over to her left, there is a flare of dark scarlet flames. Someone behind her screams —shock, Aliyah guesses. Even three years into this, the violence out there is still surprising. The woman in front starts to run. Before Aliyah knows what her own legs are doing, she sees the woman has fallen and she is running now too, running until she is a couple of metres behind.

She watches her own arm reaching out to the fallen woman and the thought in her mind is *attack.* The woman turns, catches Aliyah's eye. They both freeze.

The world seems to widen and stretch and then Aliyah remembers the feel of her own body. The weight on her own shoulders that is the notebook, stowed to come home again. The strain and numbness on her thigh, the exotic sense of heat that is spreading. The thudding of her blood in her veins. All the reasons not to be rash. *You can't risk being found out. Not now. Not when you're so close.*

The fallen woman stands, slowly. She doesn't brush the dirt off her knees. Before she turns and starts walking again, Aliyah can see she has split her lip and there is still fresh blood trickling down her chin. In the fading light, her eyes are dark and hollow. Perhaps she is vaguely familiar after all, one of the technicians from the testing lab? The woman begins walking and, after a slight pause, Aliyah does too albeit at a slightly slower pace, willing the woman to pick up speed and pull a safer distance from her.

The rest of Aliyah's walk is a blank. When she thinks she hears rustling close behind on her left, her eyes fill with tears and she keeps going with fists clenched. A lifetime passes in only a couple of minutes and then she reaches the first of the working halogens and the spire of the church and her breath catches up with her again.

As the week passes, Aliyah's exhaustion grows. She grunts at colleagues until they leave her alone, fumbles through her days until she blinks and finds it is Thursday and she is standing in the queue at the beaker station, again, moving by instinct to go home early and collapse in bed. *One more day to go.* She has stopped wondering if she should postpone the meeting with the doctors tomorrow. None of her calculations suggested it would grow at this rate. She can only assume this is a success.

When she was first accepted into the University, Aliyah had a clear picture of what success would look like. An office of her own, with a crystal decanter of sherry on the far edge of her green leather-lined desk. A tweed jacket slung over a comfortable armchair by the towering bookcases. The kind of silence that means *detached* and *residential district* and *large, private garden.* Invitations to drinks parties—what would happen there was fuzzy, because Aliyah actually disliked social events. But when she was successful perhaps that would change too. The important thing was the money and the time it would buy for her. No more

carrying shopping or cooking or cleaning. She used to listen to the podcasts about the latest mass-market upgrades and daydream about bins that took themselves out.

Finally, she is at the front of the queue. Her hands follow the cartoon instructions pasted up on the wall by the bio-wash sinks. She bends awkwardly to tie a loose shoelace, the now clearly defined dimensions of the experiment obstructing her comfort. The walk back to her studio passes without notable incident, although Aliyah does shudder slightly when she passes *that* patch of hedgerow and she turns her head to avoid seeing the raw, exposed foundations beyond it.

One slice of bread in the toaster. Aliyah picks at her cuticles and dreams of expensive hand cream. She runs her fingers down the misshapen, clumping spine of the stitches where she has done a clumsy job of tacking up the tear in her coat sleeve. She could fix it, now. Unpick and start again. But there are still those papers in her rucksack from the labs on the continent. She makes herself stop, stand still, breathe. The cracks around her window are letting in the smell of smoke and she thinks about fireworks and crowds—men and women together, children even—and food so plentiful it can be left on the ground if dropped.

Breathing in for a count of three.

Breathing out for a count of four.

In her mind, the crowds disperse. Just the bonfire left, the heart of it burning red and bright. The autumnal tang of the air and the gunpowder scent shifting like a breeze until only a rich, oily metallic smell is left. The bonfire cracks and opens, a giant maw. Above it, replacing the stars, two pinpricks of red focus on her.

The toaster pops up. Aliyah opens her eyes. The papers from the continent it is then, a welcome distraction. She flicked through them briefly at the lab earlier. Months now, she's been hoping for more efficient manual processing analysis techniques and it finally looks as though someone in Spain may have something promising.

The numbness on her inner thigh is becoming itchy, the heat receding as fast as it arrived. She presses the flat of her hand down and tries not to guess what it means. *It was the only way to check. The only way fast enough to make a difference*. Under fabric, the skin is granular; if she rubbed a pinch of it between her fingers perhaps it would dissipate into sand. Even standing still, the weight drags through her bones. The tension of her jaw a constant annoyance, her tilting pelvis making her stance as well as her mood crab-like. More efficient techniques would be helpful: easier to scale this up if it works.

If.

When Aliyah first spotted the pattern she thought it must be a mistake. Surely an answer that obvious would have been noticed? But there was nothing in any of the photocopied, stapled research papers still being distributed to suggest anyone was working on it—plenty of other theories and trials, but nothing about this kind of targeted deactivation. Nothing on the podcasts or news reports. Nobody in her team brought it up at meetings.

Synthetic DNA, produced from the original materials of the buildings. Modified using her team's techniques for targeting T-cells. Chimeric antigen receptors forming like roof tiles.

Technically, she shouldn't be giving her own blood for the department's studies, but the clinical doctors had been turning a blind eye to that for the past couple of years since the sacrifices began and people were less willing to come in and donate in case the beady eyes of the doctors spotted something was wrong with them and wrote them up as 'volunteers'. It was an easy conversation to introduce as she sat next to the apheresis machine and swung her legs. *What if we could do something really simple? On the QT? Nobody would ever have to find out if it didn't work.*

They'd said yes immediately, the pair of them helping her smuggle the extra samples out in a cool box. And she passed through The Dead Zone every morning anyway, so nobody would question her pausing to slip a handful of rubble into her pocket. She has included nothing in her research notebooks about her reasons for choosing samples from that particular vacated patch. Nothing about the gorgeous, crenellated building that used to stand there and the rich fantasy life she has been imagining for herself within it since well before any of this happened.

The first two cultures produced nothing. The third—slower than anything she'd worked on before, but there was definite replication.

Something rankles in her head. A wrinkle that won't be smoothed away. Julia—Prof. Eskbank—and her eyes travelling over the shape of her body. Assessing. Clinical. *Has she guessed?* She feels angry with the older woman, a flash that echoes unusually deep. Aliyah wants the success to be all her own. To save the world by herself and to pick up—well, not where things left off exactly. To go back to how the world was, but this time as a winner.

Besides, she's violating every ethical and safety procedure. If Julia found out, she'd be obliged to fire her. Aliyah is banking on the hope that the expediencies she has introduced to her methods will be forgiven as a *fait accompli*. She's heard too much about other trials stalling, stifled and repressed thanks to red tape.

Professor Eskbank's smile floats before her eyes. *The sick are too useful.* Half delirious from exhaustion, Aliyah wonders how much the older woman has to pay to keep the age registered on her ID safely below the threshold for compulsory redundancy and voluntary sacrifice.

The pain. The discomfort of extreme heat. The numbness—and now the itching. It will all be worth it if she is right. Synthetic DNA, mimicking the structures of the buildings that have gone feral. A new pathogen that can be weaponized, once she's proven the theory works. Injected into the nest out at the airport, perhaps. Or dropped in a smart bomb. That part will be someone else's problem. Aliyah remembers rumours of what happened when the military took over one of your projects, back in the days when there were government-funded armies still. Perhaps, she hopes, one of the private militia will think it has the kind of *potential* that leads to a big cash payoff. However it is achieved, the important thing is that the new organic data will move through the buildings' entwined systems, deactivating them.

A wild picture dances in her eyes, of moving in to one of the deactivated mansions. The Escher staircases twisted to the point of movement. The lean of it, frozen mid-step. If Aliyah—tired, weak and petulant—can get the samples tomorrow, with the doctors' help, then she will have taken a massive step towards setting the world back on the right tilt. Taking the same principles from her team's research into cancers, applying them to a new problem. Targeting the issues and eliminating them. The research itself dividing and replicating and living on in new shapes, new structures: all dormant. Taking back control of the data. Aliyah avoids thinking about the necessity of the doctors' role in all of this. The blood and the bandages. The permanent limp. Her mind drifts on to imagining how her reputation will be lauded after her experiment is removed and shared and applied more widely. How the world will shake and settle down again, refreshed.

The world *is* shaking. Aliyah crumbles the remaining crust of her toast and it drops on the floor. She grips the counter-top and watches as the toaster slides along and is held by the plugged-in flex. She reaches out automatically then stops herself, unsure if she switched it off at the wall or not. Unsure if she can bend down like that anymore without damaging it—damaging herself.

Outside the window, flares of scarlet. She can hear someone screaming and her throat is sore—she thinks the room is full of smoke then blinks and realises it is not, her throat is just sore because it's her who is screaming.

The toaster strains against the plug and falls. The building shakes again. Aliyah falls too, and one of the cabinets falls and bangs her in the stomach and then again on the thigh. She screams again, this time in sheer pain and then again in fear because she needs to live through this undamaged. She needs to make it in to the lab tomorrow. As the roaring outside intensifies she is reminded of a sunny afternoon when she was new to the University and one of her fellow students—a man, one of the posh ones who was first to go—parroting a phrase they had just heard for the first time at a seminar. *Publish or perish.*

She pulls herself along the floor to the window. Her tenement is still now. Aliyah forces herself to look out, as she does every time there is a visit. There they are, lighting up the horizon.

The same principles, she reminds herself. *If it can be grown, it can be targeted.*

The group of buildings move off. These ones are modern on the outside as well as the inside, none of the crumbling Georgian sandstone of the buildings from The Dead Zone she has to cross to get to the lab every day. These ones must have come from the developments on The Shore. Their sleek glass twists and reflects the moonlight as they lurch along.

Aliyah didn't know many people who died in the first wave, when the data-enriched buildings of the successful, the powerful and the financially comfortable unlocked themselves from their foundations and began to display an unexpected hunger for human tissue. Nobody *she* knew well lived in houses like that. But she recognized the names in the news reports—all the senior management from her labs, all the politicians. Then what remained of the government swept in and closed the labs that relied on connective technology to do their research. Called them *super spreaders*. She remembers the hatred on the faces of the people picketing the edge of The Dead Zone, the handwritten notes pushed through her old letterbox smeared with dogshit. The advice from what was left of the University's admin team to tell her new neighbours she was a nurse, not a scientist.

'We're fortunate we're behind the times.' Professor Eskbank handed out oranges when the team straggled together a few weeks later, reminding them of the importance of maintaining a varied diet. 'I've printed what I could. I know most of you are late submitting your figures—I assume you've had the sense to bring your backups with you today.' The message was clear: they could continue working, even while the richer parts of the city crumbled into motion. 'We will be needed,' Professor Eskbank said, finally running out of fruit. 'Until there is nobody left to benefit from our research.'

Her thigh moves again, a noticeable judder across the grey, exposed flesh. A tug from the sinews inside. Aliyah pulls a rug over it then, suddenly afraid, props a blackout cover against the window as well. She lies down on the pull-out bed and thinks she will never be able to sleep but she does, the fast and undreaming descent of the exhausted.

She is halfway through The Dead Zone when her leg stops working. There's nobody else in sight, just the blood-red dawn lining what used to be a historic skyline.

Stupid with tiredness, Aliyah thinks for a moment that this is what she wanted: for the buildings to stop moving. The structure that was her leg has rooted and stuck and this is fine because it's the point of the experiment, surely? Then she wakes up enough to think about what will happen if she is caught out here. The questions that will be asked. The waste of it.

As she looks down to the unmoving limb, she sees from the bulge under her trousers that it has spread below the calf now. *I should write that down in my notebook.* She puts her hands there—there's no avoiding touching *it*, no bending low enough to touch just her leg—and heaves.

The leg moves forwards and drops again. Aliyah repeats the almost-pointless shuffling for another few metres until sheer frustration allows her to find the strength to swing the leg. She falls into a regular rhythm—swing, step, swing, step—and is past the empty road outside the Ashworth and making her way round to the side entrance before she realizes she hasn't seen a single building between the start of The Dead Zone and the labs, moving or inert. *As if the last twenty or thirty have been stolen overnight.* Her whole skin prickles—not just the itching lump of the experiment—and she wonders if she should tell someone.

The lab is completely empty, even though she's running late. Aliyah is so unnerved by the journey that she doesn't bother going to prop herself in a toilet cubicle. She part-strips by her workbench and there on the growth are what she can only describe as windows. A deeper hollow that might be a door. She wipes her running nose and scrunches her eyes to peer more closely under the harsh strip lights. Is that a faint glow of red through the skin, or is she really losing it now?

Her fingers tap an unconscious pulse against the workbench and—she really doesn't want to admit it to herself—her thigh pivots to face the tapping and *sniffs*.

Any remaining thoughts Aliyah has been harbouring about putting this off dissipate. She wants it out *now*. She starts to dress with more haste than speed, all the human parts of her trembling and the experiment—this thing that used to be her own leg—mute and solid and motionless again. The door of the lab opens and she turns to greet the doctors, relief at seeing another human being making her willing to overlook their lack of punctuality and her own state of undress.

'Good morning Aliyah.' Professor Eskbank stands uncomfortably close, poker straight and with her mouth set in an unwelcome smile. She pushes Aliyah back down onto her stool with a tap of her fingers and leans against the workbench beside her. 'A pity. You were quite promising. I didn't think you were such a complete idiot.'

'I didn't think it would spread so fast.' She starts to offer an explanation, but Professor Eskbank raises a hand.

'May I?'

Aliyah nods. Her boss takes one of the blades from the shelf and slits the trouser leg from the bottom to just above the knee, taking care not to touch the flesh, the experiment, itself.

'Beautiful, in a way, wouldn't you say? But I really will be sorry to lose you.'

Perhaps, Aliyah wonders, part of her relief is that she will no longer have to cross The Dead Zone twice a day. No longer lie to her neighbours about where she works. No longer spend her working hours feeding cell cultures to find a cure for diseases that no longer seem important because the sick—the terminally ill and those who pose the greatest strain on resources, the latter pressurized into volunteering *for the greater good*—are fed to the buildings.

No more being driven to desperation. To experiment. To push back at the architectures of flesh in the same way architecture itself was originally modified to respond to systems of data pulsing through it.

Professor Eskbank snaps her fingers in front of her face again. 'I said pay attention, Aliyah. Do you know what happens next?'

'The doctors. They know. They're going to help me get the samples. Get it out. I think—' Aliyah swallows. She doesn't want to admit this has begun to spiral out of her control. 'I don't think I'm going to be able to keep my leg. We have to amputate.'

There is a shocked silence. 'Oh, Aliyah,' Professor Eskbank looks genuinely sorry for a moment, before her usual cynical scowl slips back into place. 'Oh, you really—the doctors aren't coming. Nobody is coming. This is through your whole system, do you understand? I can't cut this out of you. You have to be alive, do you understand? If there's a chance this is going to work—and I've been looking at the notes you passed on to them last week. Very interesting. You might be onto something. It's certainly worth testing, but I can't risk seeing how it transfers. You'll have to go in alive, in the truck with the other volunteers.' Her eyes search Aliyah's face, waiting for comprehension that does not rise up. Her tone hardens. 'What did you think was going to happen?'

'A vaccine—trials. Something that could stop them. Something for the buildings to ingest.' Finally, Aliyah realizes what her boss means. 'No. I can't—I'm not willing to. You can't make me. We can keep samples alive. We can grow them.'

Her boss stands up and steps back, dusting grey powder from her fingers. Dimly, Aliyah realizes she has removed a section of the experiment. 'You didn't feel that? Good. Total inertia. As predicted. Faster than in my own tests. I will be fascinated to get a proper look at these.' She holds up Aliyah's notebooks, a loose sheet flutters down to the floor between them. 'Not now, of course. There's no time now. We need to get you out there before this takes you over entirely. Haven't you seen the news, Aliyah? Time is of the essence.'

Phrases Julia has let loose at their meetings slip through Aliyah's mind. *The sick are too useful*. Her leg tugs again, but the rest of her can't do anything about it. Now it is she who is too heavy for her leg to move, not the other way round, and she wonders if anyone will ever find out and laugh on her behalf at the irony. Her boss runs a finger idly across the shelf above her workbench and tuts at the poor housekeeping, glancing occasionally at the clock on the wall.

Finally Julia gives Aliyah a brisk nod and says goodbye. 'Someone will be along shortly to collect you. I really hope this is it, Aliyah—I hope it's you who makes a difference. It must be such a comfort, to know you are a part of something greater. That your work will live on as a touchstone of a salvaged civilization.'

'Under whose name?'

Julia gives a small smile and leaves, nothing but a faint scent of clove and spice. Aliyah closes her eyes and wonders how it will feel. Whether she'll still be conscious when she's dumped out the truck. *It takes longer for them to spot you if you don't run*, or so the rumours go. *It takes longer for them to track you if they can't sense your pulse.*

Footsteps in the corridor. Under her unmoving thumb, something in her wound shifts.

A Note on the Science
by Sonja Prade

The gene-editing technology known as CRISPR (clustered regularly interspaced short palindromic repeats) is now frequently used for optimising therapies and diagnostic tools in the clinic. It was originally found in bacteria that use it as a defence mechanism against invading viruses. Scientists have learned how to repurpose it to edit specific sequences in human DNA. CRISPR technology has potential in cancer therapy, but also in gene therapy by introducing protective or therapeutic mutations in host tissues. It is a cost-effective and convenient tool for various genome editing purposes. For instance, human T cells—immune cells that are central to our body's defence system—can be modified to become more efficient in recognising and fighting cancer cells. Deleting genes that are known to inhibit T cell function can improve their response against cancer cells.

Viccy Adams is a short fiction writer based in Edinburgh, with a background in creative research. She was Leverhulme Writer in Residence at the School of Informatics, University of Edinburgh, during which she edited an anthology of mixed-media work about Alan Turing. She has been a writer in residence with the Naxi people in rural South West China, created a virtual library of books by women, had writing about sewing machines exhibited at the V&A, and has burnt a day's work to fuel a mobile sauna. Her non-fiction collection, *Recollections: 12 vignettes from Lashahai*, was shortlisted for the New Media Writing Prize and won the People's Choice award. Recent commissions include *Spree*, a digital narrative experience for a 14+ readership as part of Nesta's Alternarratives strand.

Dr Sonja Prade, is an immunologist working on cancer immunotherapy. She gained first insights into the field of tumour immunotherapy when studying at the Ludwig-Maximilians-Universität in Munich, where she completed her PhD focusing on patients suffering from autoimmune diseases. Her current work at the University of Edinburgh combines cancer immunology and autoimmunity. Cancer cells have several mechanisms to escape recognition by the immune system, rendering many therapeutic approaches unsuccessful. However, recent advances have been made using T cells to target cancer cells with some remarkable responses. T cells can be equipped with molecules to specifically recognise cancer cells, and using T cells with autoimmune traits can improve their function. These 'engineered' T cells can be more effective in controlling cancer growth and help overcome the limitations of today's immunotherapy in the clinic.

Pippa Goldschmidt

Branching Out

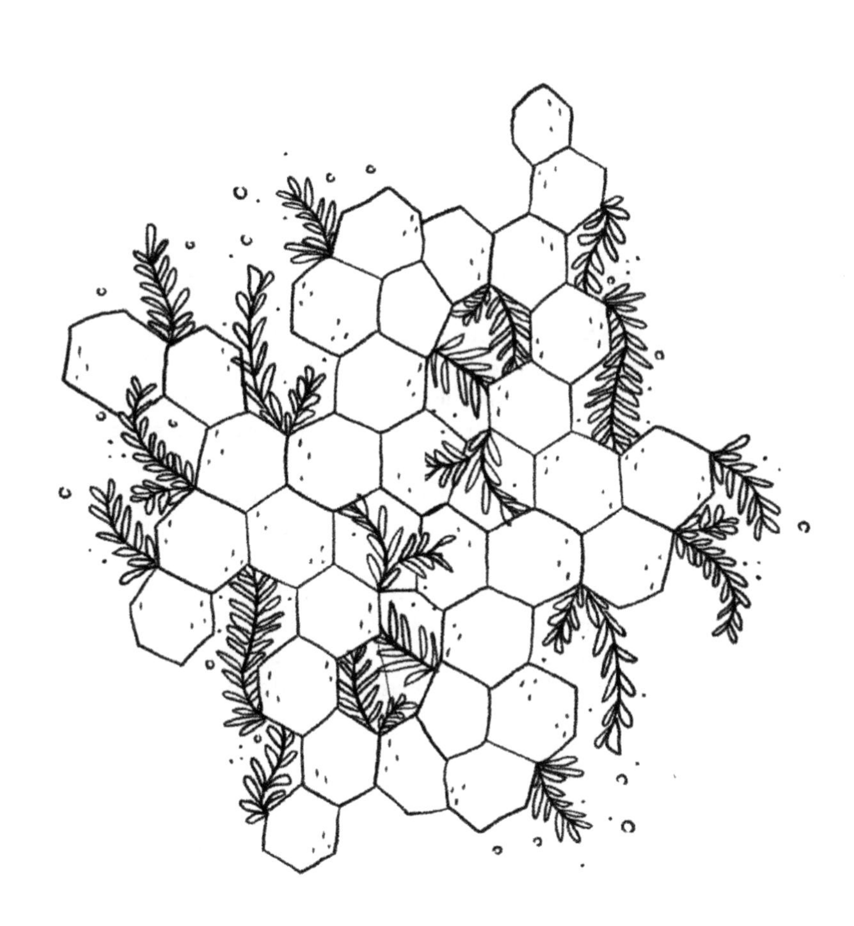

You swoop over St Magnus in the heart of the city, close enough to see where the surface of the cathedral's tower has worn away in coin-sized pockets and needs repair. You fly above the Stones of Stenness, able to identify individual fronds of mustard-yellow lichen growing on the rock, an accurate measure of air quality. You travel along the A965 towards Stromness and the new builds, assessing the degree of disturbance to the surrounding moorland—

A colleague pops her head around the door, 'Tea?' You nod and return to the footage. You continue to measure, check, process and compare, before you input the data and the related analysis to the Vault. Tea arrives and you pause to exchange pleasantries with the rest of the team. It's most important to take regular breaks from this demanding and important work. Someone hands round slices of homemade cake. You finish for the day and head home.

*

I'm heading home when I see the fox. Trotting purposefully along the street, its nose scenting rubbish, zigzagging back and forth across the pavement, and only slowing down when it reaches the café opposite St Magnus. Just an ordinary fox, doing what foxes always do in cities, and at first it doesn't register. But I've never seen a fox here. They don't live on these islands, they never have. How did it get here—by hitching a lift on a boat across the Pentland Firth?

The fox pauses in its search for food and looks at me before it jinks off down one of the dark vennels as if it knows exactly where to go.

Later, lying in bed, I conjure up all the separate parts of the fox that I can remember. Whiskers, snout, a tail slung low. A walk that can best be described as swaggering. *You act as if you own the place.* I sit up in the darkness; I haven't heard that voice for a long time. I don't want to hear it ever again.

*

When you first arrived on the islands you had a job on the seaweed farm. You were invited to move up here from your old location because of your experience with robots and your ability to dive. You were allocated an office in the old council building at Scapa Flow. You were also allocated a wetsuit, the job required you to climb into it at a moment's notice and sort out problems with the mini submersibles. The first generations of subs were apt to get tangled up with the seaweed and you had to cut them free. The new models are much more efficient, and have been trained through a process of machine-learning to both seed and harvest the seaweed with little or no human intervention. You stayed in this job for some time. The wetsuit got less and less use.

*

The wetsuit hung on the back of the door like a skin like the abandoned fur of a seal. I thought I saw a selkie once, when I was out diving around the seaweed and it was swimming closer to the tidal turbines' blades than any of us would dare. Look at it one way and it could have been a woman, blink and it turned back into a seal. There are always two sides to every story, but I swear it looked as real as that fox.

*

At work the next day, you all tell each other about your different sightings of the fox and this triggers a timely discussion in the office about the problem of invasive species. Animals and plants threatening a fragile ecology that needs protection from the outside world. It's an ongoing problem, one that requires constant invigilation. Everyone knows this is the only way we can make progress here.

When you have eaten your lunch you are asked to monitor the growth of the slime mould as it creeps along the latest and most detailed model. The slime mould is a biological computer, able to calculate the shortest routes in a complex terrain of natural and artificial entities, and these routes will be used to decide the locations of future houses and other buildings. This is the first city in the world to use such innovative ways of solving the problem of developing an ideal city for its citizens. A virtuous circle of data-gathering and planning, and all within the Vault. Already the Vault has shown itself capable of planning energy generation and consumption in a more sophisticated way than anywhere else. Because of the precise manner in which the Biomasser is programmed to generate electricity, citizens find their houses are warmer in winter and at a lower cost than elsewhere. Thus is the city self-contained and able to withstand energy shortages that are crippling other cities around the world.

*

After work and it's already dark. I find my way home by the low-level lighting that anticipates my route through the older quarters of the city. As I turn left and right through the maze of the vennels, lights set into the paving slabs starts to glow, always a few steps ahead of where I want to go. The slime mould calculated this route as the most energy-efficient way of guiding me home.

Near the cathedral two eyes glint at me and I wonder what the fox is eating. Rubbish? Endangered indigenous mammals? Does it feel continual amazement in this new place or perhaps it's just lonely. *Own the place*. No. Above me, a diffuse glimmer in the sky from the Biomasser at Scapa, burning the seaweed. Turning it into heat that is converted into electricity that supports the Vault that plans the city. It sounds like a nursery rhyme. *What's the time, Mr Fox?*

*

In the footage of the dead fox, the images are detailed enough for you to notice the bullet hole through its forehead. You add the footage to the Vault, inform the team that this latest instance of an invasive species has been solved, and you drink your tea.

The team receives an invitation to visit the new builds and see how the latest extension to the city is progressing. You accept the invitation. New ways of building houses are being tested here all the time. They're so interesting to learn about.

After lunch and you're tasked with setting up another slime mould enquiry. This new generation of cyberspace mould is proving much more robust than the original one, the physical *Physarum polycephalum*. Although every care was taken with its housing in Perspex boxes that met the most rigorous biosecurity standards, nevertheless the slime mould could not cope with the incursion of an invasive species. It is still not known what caused the presence of this species in the boxes, or indeed anywhere on the islands, in spite of extensive investigations. Cyberspace mould is as accurate as physical mould and is now protected from hacking by sophisticated anti-malware that can always, no matter who claims the contrary, identify the location of the would-be hacker with pinpoint precision.

*

This evening I should be going to my Norn evening class but it's cold and dark out and I'd rather just have a bath. At first I was keen, perhaps too keen, to fit in. Straining to shed some essential layer of myself, the way a selkie discards its fur when it abandons the sea to become a human.

After the bath I feel a bit more invigorated and I decide to tackle the boxes again. Since I moved here I've never bothered to open all of them and they've sat in the spare room ever since. They're full of books as well as jumpers because I was warned about the cold Orkney winters, but I've never found them that bad.

You. Act. I kneel in front of the boxes. *Own the place.* Time to unpack everything. I need to be more organised, like I am at work. I've earned the right to feel at home. So why don't I? Is it because of the farmer?

I rip open one of the boxes but I use too much force and everything spills out. As I expected, many jumpers and many more books. A mobile phone that was already obsolete when I moved here. Some postcards bought in art galleries back home. A sealed envelope with no name or address written on the front, so I open it. Nothing inside it either, and I'm about to crumple it up and throw it away when I spot something tiny in the crevices. I shake the envelope and a shower of black seeds cascades into my hand and onto the floor. Seeds? I have no recollection of dropping seeds into an envelope and bringing them with me on that long and tedious journey north all those years ago.

Earlier today, the fox's fur shining so bright on the screen that a tear came to my eye. The bullet hole clean in the middle of its head looking almost incidental, far too small to have killed an entire animal. Even as it lay there dead in the field, its whiskers and snout were just as I remembered. Still, I sent it to the Vault. Not much of a burial but it'll have to do. The fox will take its place in the city, because I'm beginning to realise that in some ways the city of data in the Vault is more real, certainly more complete than the city in real life.

I recognised where the fox had died, because the field was where I encountered the farmer. It was one of the larger plots on the hill behind the Grammar school. I'd gone there keen to see how the landcrops actually grew in this place because, until I moved here, reading all the manuals and papers felt a bit too theoretical. I wanted to see for myself the experiments with vernalised wheat and so I was standing near the edge of the plot, crouching down and running my hands through the tender new growth, and appreciating the pale greenness of the stalks when I heard someone shouting. A man, somewhat red in the face, and with a shotgun broken over one arm. I stood up and carefully explained who I was and why I was interested in the crops. I told the man that, before I'd even arrived here, I'd read all about the experiments on the seeds to try and make them germinate in cold weather in a place where, traditionally, wheat had failed. As I talked, I was still hopeful that he might be listening. When I finished, there was a moment's silence and, stupidly, I thought it'd be all right, until he leaned too close to me and whispered, 'You act like you own the place.'

There was a pause. I couldn't think of anything to say.

Then he added, 'Your sort always do.'

Your sort.

I reached down and plucked some wheat from near my feet. I held the stalks out to him so he could see how I'd severed them from their roots before I put them away in my pocket. 'So interesting,' I said, 'did you know that the Soviet Union experimented for years with vernalisation?' and I couldn't stop myself from continuing, 'The farmers all starved to death.'

I watched him stroke the barrel of the shotgun as if it was alive. A distant bird trilled, a two-note warning song that I should have paid more attention to.

'Did *you* know, Missy, long after hunting with guns was made illegal in other places, we just carried on up here because nobody was checking up on us, nobody gave a damn about what we did.' By now he'd straightened the shotgun and it was ready for action, although still pointing down. Nearby, a hawk was hovering almost immobile, scanning the ground. I remember wondering if it had noticed us and if so, what it thought we were. Prey? The gun glinting in the afternoon light suddenly wheeled around, and before I could understand what was happening, a blast shattered the space around us and the hawk plummeted out of the sky and down into the wheat.

He broke the gun again and it hung obediently over his arm. I was immobile with shock. Not just from the killing of the beautiful bird but also from the transformation of the afternoon: a hole had been punched right through it so that I couldn't join up *before* and *after*. Couldn't make sense of the events, they were like a series of images with some essential bits erased, or redacted.

'Now you, Missy, or Doctor, or whatever the hell you call yourself, ask yourself this question. Why am I able to do this, without anyone noticing? Without anyone in the Vault—' here he paused before continuing, 'seeing what happened, when you think you can see everything? Just like that hawk?'

Another gap in the timeline; I have no memory of how I got home from that field. But he was right; when I checked the next day I couldn't find any footage of the violence in the field. All I had was a handful of limp wheat shoots, which I left on the kitchen table until they withered away completely and then I buried them in the compost.

Now, I gather the small black seeds and carefully replace them in the envelope.

*

The new builds are all coming along very nicely. You're issued with hard hats and given a guided tour of the site. You're shown where the heatpump will go, where the raised beds and greenhouses (to be warmed by the heatpump) will go, as well as the charging points for the electric vehicles. You follow the manager of the project as she leads you around and explains the function of every aspect of this settlement. You've seen it already of course, you watched as the slime mould identified the optimal spacing for the houses, for the paths to the main road, for the siting of the raised beds. Out here, in the sharp wind coming off the loch you can only admire how the slime mould has anticipated every aspect of the future lives to be lived in this place. A perfect symbiosis between inhabitants and environment.

Crops will be specially chosen to suit the microclimate. Outside spaces are being designed to encourage people to do more physical activity that will improve their health, which in turn will decrease the need for expensive medical services. The current generation of agri-robots will be adapted to work in this site. Everything is planned; nothing will be left to chance. You all have a most instructive afternoon that will enable you to do your tasks even more efficiently back in the office, now you have up-to-date first hand experience of the goals of the city.

*

I must admit that I was briefly tempted to scatter some of my seeds in the earth around these new builds. Just as an experiment, to see how this would affect the planning.

I don't know what these seeds are, I can't remember. At the time it must have been so obvious that I didn't even bother writing their name on the envelope, but I've been here too long. Now, the only way to identify these seeds is to grow them.

Of course, we're encouraged to grow our own plants. The city flourishes in the spring with window boxes and flowers pots everywhere, as well as the smaller patches of harvestable land here and there. We all belong to a gardening club which hands out advice and tips and which judges our efforts at the annual show. The green in the front of the cathedral has, for many years, been planted with strawberries that anyone can help themselves to. A gift from the city. A communal greenhouse sits on the runway at the old airport. Raspberry canes twirl up the side of the town hall, an apple orchard has spread towards Ward Hill. South, between Kirkwall and Scapa, are the larger fields of barley.

Now, I take a plastic pot from under the kitchen sink and fill it with soil from another houseplant. Something tells me not to go

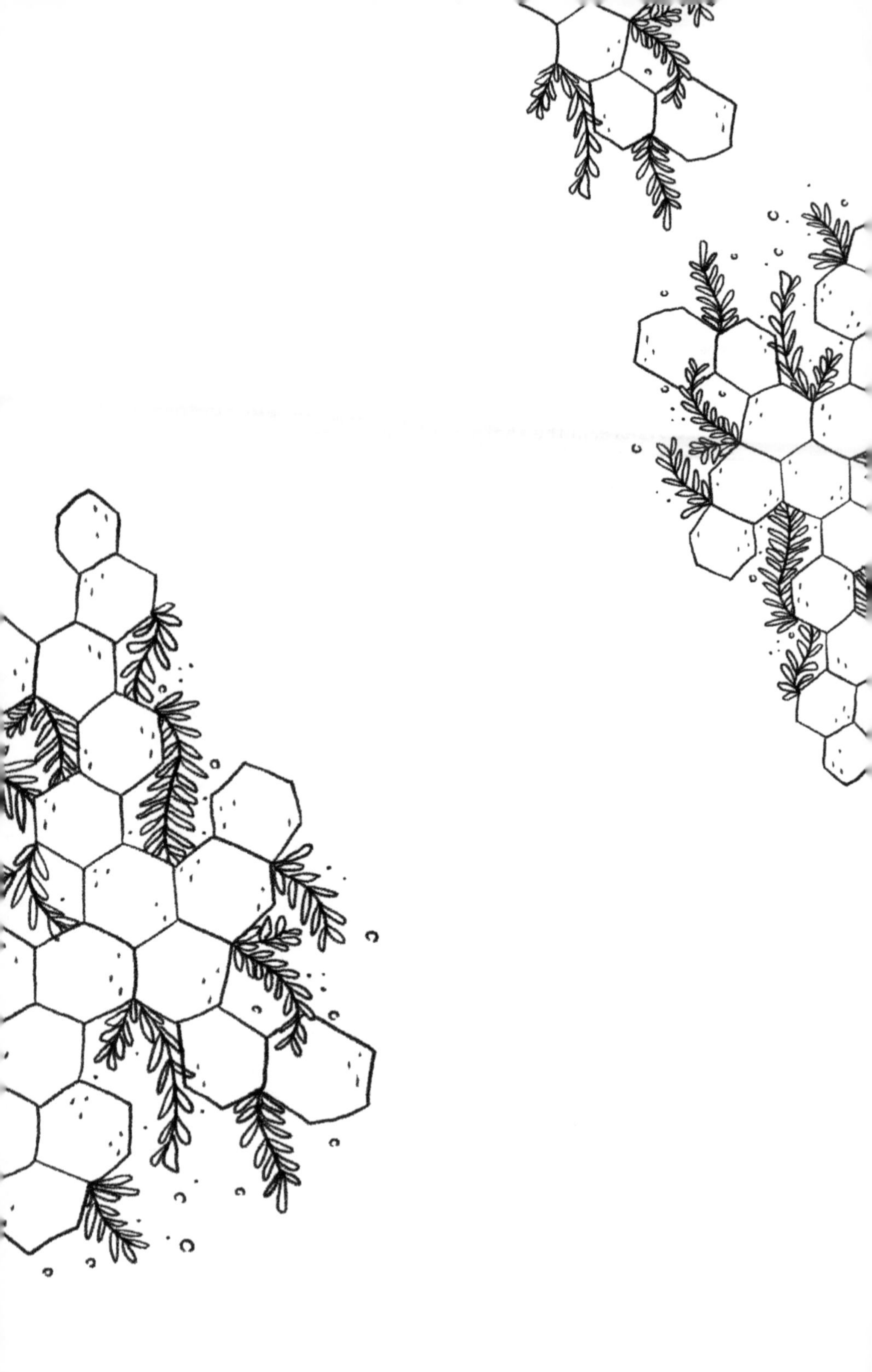

outside and get soil from the garden even though that would be the obvious choice. Something tells me to do all this inside.

I sprinkle the seeds on the soil and water them. And I settle down to wait.

*

Whenever you have a quiet moment at work, you play a little game with the Vault. Choose a position on the map of the city and zoom in. Houses and streets appear. Zoom in closer. See the individual tiles on the roofs, the paving slabs on the pavement. Zoom in closer. See the grass growing between the cracks of the paving slabs, and the blossom on the trees. Zoom in closer. See the birds and insects in mid-flight. Zoom in closer. Keep going all the way, admiring the incredible image quality. Only then can you appreciate how much data goes into the Vault.

Sit back, rub your eyes. Relax. So many images. So much data in the Vault that plans the city that in turn supports the Vault. A fully virtuous circle with nothing and nobody left out.

In the next extension to the city, the Vault has designed a building with a honeycomb structure full of hexagonal small cells, each of them branching off from a central portal. The design is very clever, each cell can be viewed from the central portal but no cell can be viewed by any other.

You ask the Vault why the city needs this building and what function it will have, but the question is identified as an error and no answer can be issued.

*

I keep my pot planted with seeds in the middle of the kitchen and well away from the window. This seems sensible although I'm not exactly sure why. It's not an ideal position, of course. When the seeds germinate and the first seedlings appear they look even more spindly and delicate than normal. I turn the pot every day so that all the seedlings receive the same amount of sunlight, and they bend towards it, thirsty for it.

The pot starts to teem with these long and delicate seedlings and I still don't know what they are. I can't recognise them. But they're obviously not healthy; they're competing for limited resources. I try and thin them out so that the surviving plants don't have so much competition and I put the discarded stalks in my compost heap, covering them with vegetable peelings.

*

The Vault tells the managers of the office that they will need to recruit more people to work here, to process the ever more detailed data and manage the slime mould as it makes ever more complex enquiries.

Part of the essential plans for the city is an expansion of the Vault itself. The city can't function without it. The old distillery near Scapa will be repurposed for this expansion, which will need several hectares of land for all its databanks, as well as cooling from seawater.

You will need to monitor very carefully the land set aside for the expansion to the Vault, to check for anything out of place. The build is scheduled to start soon and you're tasked with checking the footage. It's a promotion and your colleagues are delighted, someone else brings a cake into the office for an impromptu celebration. It's because you're so good at your job, the cake-bringer says, perhaps you have a greater ability to concentrate than the rest of us.

The mould will be tasked with mapping the best route between the Vault and the city. The mould has proved that it will always solve the problem of getting from A to B, no matter how complex the route is.

*

When I leave the office, it's late and already dark. I'm walking along Albert Street when I decide at the last minute to take a right turn instead of my usual left. The way ahead of me remains dark, the low level lighting doesn't kick in, the city fails to recognise my journey. Although I can barely see where I'm going, I continue walking in darkness. To guide myself home I have to run my hands along the stone walls that enclose the vennels, but I don't give up.

Once I'm home, I hurry to the kitchen to check on the plant. But its leaves are too pale and I have to support the surviving stalks with chopsticks and pencils to stop them flopping on the table. I still don't know the name of this plant, I could take an image and look it up online but that doesn't seem wise.

As I eat dinner I look at an old paper map, published when there were still fifteen miles of open land between the city and Stromness. I do a quick calculation: the new Vault will be capable of storing petabytes of data. Each aspect of the city will be mapped, down to micron level. The city in the Vault will be more understood, more analysed than the real-life version. In some essential way, the virtual city will be larger than the physical one.

*

At work you are asked to check the slime mould's most recent enquiry. It has been tasked with re-calculating routes through the part of old city near where you live. It finishes this task and displays the results: the reworked route cuts through your house, which no longer exists on the screen. The mould has done this enquiry in record time; it's been proved to be a more efficient computer than any digital equivalent.

*

In the kitchen. The plant is about a foot tall now, and the surviving strands have somehow grown together into a sort of plaited vine, pale leaves sprouting out at odd angles. I have never seen anything like it. I certainly don't remember any similar sort of plant living in my old home down south. Although it still looks too pale and sickly to survive, when I press the vine, it feels surprisingly resistant. The pot is too small though, so I take another larger plant and unhouse it, so that the vine has a new home. More space to grow.

The plant seems the apparent opposite of the slime mould, which is virtual and neat and sees all the data. This plant in my kitchen is real and blind, and somehow chaotic.

*

The Vault says it wants to talk to you about your house. The Vault tells you that it has noticed serious structural problems in the house. You don't reply. The Vault asks you if the kitchen lighting is satisfactory. You don't reply. The Vault records your silence. It states that you cannot deny an official entrance to your premises in these circumstances.

*

At home, I sit in the dark kitchen. I find myself talking to the plant again, which is flourishing since I repotted it, and appears to have adapted to the low levels of light in the kitchen.

There is a knock on the door.

A Note on the Science
by Karen Halliday

Growth and behaviour of living organisms are strongly influenced by external stimuli. These features are easy to study in plants as they are exquisitely sensitive to changes in the light environment. In low light, plants often develop a pale, spindly architecture. But appearances can be deceptive, the spindly growth format is the physical manifestation of an adaptive response that is critical for plant survival in low light. Plants can have remarkable powers of growth, plasticity and agility, and so they offer a portal to understand how the external environment can shape development. By studying molecular, cellular, organ and whole plant responses, scientists are learning how biology is configured to react to external cues. Synthetic molecular tools, 3D imaging and modelling approaches are enabling scientists to build a dynamic picture of how plants respond to light changes in real time.

Pippa Goldschmidt is a writer based in Edinburgh and Frankfurt. She's the author of the novel *The Falling Sky* and the short story collection *The Need for Better Regulation of Outer Space*, as well as co-editor (with Tania Hershman) of *I Am Because You Are*, an anthology celebrating general relativity. Her work has been broadcast on BBC Radio 4 and published in a variety of places including most recently *Litro*, *Mslexia*, and the *Times Literary Supplement*, as well as in anthologies such as *A Year of Scottish Poems* (Macmillan), *Multiverse* (Shoreline of Infinity) and *Best American Science and Nature Writing 2014* (Houghton Mifflin). Her latest project is co-editing (with Drs Gill Haddow and Fadhila Mazanderani) *Uncanny Bodies*, an anthology of literature and essays inspired by Freud, cyborgs and the history of Edinburgh, to be published by Luna Press in 2020.

Professor Karen Halliday is Chair of Systems Physiology at Edinburgh University, with expertise in environmental signal integration, molecular genetics and dynamical mathematical modelling in the model plant Arabidopsis thaliana. She first became interested in plants after visiting the Amazon rainforest in 1991. During her PhD with Prof. Whitelam at Leicester University, she studied the molecular pathways that sense light signals and trigger physiological responses. She started her own lab in 2000 with a Lectureship at Bristol University where she studied light and hormonal signal integration. Following a move to Edinburgh University as a Senior Lecturer (2004), Reader (2011) and Professor (2015) her research scope broadened to include light-temperature interactions and mathematical modelling. Prof. Halliday's lab continues to work on signal convergence, striving to determine how the photoreceptor pathways control carbon resource partitioning. This is important as phytochrome light receptors are major regulators of resource allocation to biomass in field crops.

Neil Williamson

Mudlarking

I've been following Doug Hanlon all the way down from Central Reclamation. Navigating off the sweepway and into Glasgow's Victoriana heart, where the old sandstone is as gritty under the blazing sun as the ruins of Egypt, he isn't hard to track even at a discreet distance. As a resource agent he's got permission to go anywhere he wants, so he's making no special effort to disguise his route. As a fellow resource agent I should have no reason to question his actions.

And yet, here I am in a clandestine community car, sweating as its aircon struggles with the morning heat, and from guilt too. He's been a good workmate and a decent friend, has Doug. He helped out with the premium to get Mum a good placement when we moved her into the Heights. But now I can't stop thinking about that. We're on the same grade and I've barely a bean to spare, so where does Doug Hanlon get that kind of disposable income? And, when he takes off on his own like he has so often recently, where does he go?

Doug's Ministry of Resources van veers west and my heart sinks at the thought of rejoining the sweepway to cross the Kingston Bridge. That venerable span's entire blighted life has been one of bolstering and shoring, making its service last way beyond its designers' expectations. The authorities claim it's safe but I've seen chunks of concrete crumble into the Clyde with my own eyes. Every time my vehicle autolocks into a gap between the rumbling truck trains to make the crossing I offer up a prayer. But I can breathe easy today. The van's passing over the sweep, cutting instead through the agriblock at Charing Cross and arrowing into the West End.

The question is why? As my car follows between the walls of condensation-misted, algae-stained glass, my unease bubbles. Aside from Doug's generosity, what am I basing this on? A feeling that something about him has changed and...that thing he said when he helped me settle Mum in: 'Do you think it's fair, all this?'

My phone rings and I sigh. 'Hi, Mum.' I inflate my voice with cheer. 'How's the view this morning?'

'Aye, it's all right, hen,' my mother says dourly. 'Same as it is every morning.' I know she's never really understood that most of the residents in the cubic monolith that we call the Heights live in internal apartments and only have screens instead of a real view, but I can't help hearing ingratitude in her casual dismissal. 'I just want to get back to my own wee flat,' she says, like she always says. 'You can't open a window here. You told me it was just supposed to be temporary.'

She's right, I said that. Her neighbourhood was a flood zone and Housing had been on the verge of sending in an extraction team. Better a white lie than the indignity of being dragged out of her bed at the crack of dawn. She has somewhere to live that's safe and resource responsible now, no longer a material drain but a net contributor to the system. The damp-ridden Govan tenement I grew up in is gone. Its grey blocks repurposed as a breakwater to keep the hungry river at bay.

'So, I'll bring your messages up after my shift,' I say, changing the subject. 'Anything you want in particular?'

'Mince,' she replies, laser focused at the mention of food. 'Real mince, mind. Not the other stuff.'

'I always get you real mince,' I placate by reflex. Another white lie. No one has had real mince in decades but with the dementia eroding a little more of her every day I no longer know if she understands that or not. 'After work, okay?'

'Aye, you're a good lass, Lynda,' she says. Then, out of the blue: 'You know, I wish you'd settled down and had a family. There's more to life than work, hen.'

My breath catches in the car's humid air. 'Got to go, Mum,' I say. 'I'll see you later.'

The car slips along Dumbarton Road and through Partick Cross and I find myself flanked by tenements. Many here are still inhabited; still have privately owned bars and shops below them. For now. A mile to the south they'd have been claimed by the river already.

They're a resource agent's nightmare, these buildings. Draughty energy sinks, burdened with plumbing and wiring from the last century, if not the one before that. Even after modification they offer piss-poor return on resource consumption back to the system. As a result, the residents get the bare minimum of amenities but can you winkle them out? Can you hell. Mum's generation are blinded by affection for these buildings but, even raised in one, I'm not so afflicted. I was jealous of my friends who grew up in modern accommodation blocks where every ounce and erg of their existence was monitored and measured so that the city's resources could be apportioned efficiently. I burned with guilt for living a wasteful life. In today's Britain, we no longer have the *luxury* of waste. Two or three generations ago recycling was a choice for the virtuous to make when it suited them. These days, it's survival. I applied for the Ministry of Resources straight out of school, eager to atone for my eighteen years of tenement living. They're the tombstones of yesteryear, these places. Flood defences are all they're good for.

At the end of Dumbarton Road, Doug pulls over underneath the westbound sweepway dividing the Whiteinch marshes from the basin that formed after the Clyde tunnel flooded. I instruct my own vehicle to stop a few streets short. Watching Doug unload the crawlers from the van and lead them through the gate in the biomesh fence like a pack of hounds eager for the hunt, I commandeer a nearby monitoring drone and send the tiny machine soaring into the sky. The hovering device rotates, orienting itself against the layout of the city as it waits for instructions. The view patched to my phone shows the swollen river teeming down from the east, bullying the supports of the bridges as it flows by; the scant remains of old Govan standing to the south and, beyond, the new Ministry habitats; the belligerent river again to the west, slowing and widening, boastful of the chunks it has claimed from Renfrew on one side and Clydebank on the other; and finally to the north, the fairy tale graveyard of the West End.

In the high distance, the gleaming edifice of the Heights dwarfs everything else. It's two decades old but it remains a thing of wonder, our first attempt at a truly self-contained mass living space. The roof has room for both a solar farm and an agriblock. The basement a full scale reclamation centre, scrubbing the air, purifying the water and turning the building's organic waste into biomass energy or fertilizer or feed for the array of clever, tailored bacteria that winkle out treasure from our inorganics.

This is the piece of simple genius that makes it all possible. We are a poor country now, but one nevertheless hung up on the old nail of national pride. Everything from base metals to rare earth elements are dwindling resources on a global scale so, rather than pay a premium to China or Australia for them, we remediate them out of our junk. This treasure goes to the state manufactories to be used in new generations of British-made devices, and the value of the reclaimed materials is subtracted from the residents' resource debt. Recycling at the atomic level, making every gram count. It's a beautiful system, all of us playing our part, living the most efficient lives we can so that we can all live.

What's not fair about that? What isn't fair, Doug, is selling resources for private gain. Profiteering is a serious crime.

I instruct the drone to look for human activity in the marshes and the view spins dizzyingly over a vista of trees and shrubs, shivering thickets, explosions of fern. The picture on my British-made phone is slightly fuzzy and tinted green, but it's good enough to spot Doug's passage through the undergrowth. He's moving east through the neglected riverland south of Victoria Park. This used to be a mixed use area, industrial yards and small business units, student flats and pre-fab housing. When the river rose, most of it was abandoned, not even considered worth the expense of levelling if the water was going to take it within another ten, twenty years anyway. There are pockets around the city like this, bad land left return to wilderness. Certainly, this place has become a jungle. All that can be seen of what used to be here is a single row of neglected tenements close to the sweepway.

The drone swings away from Doug to focus on someone else. A child of eight or nine close to the river's edge where the undergrowth is sparser, poking at the mud with a stick. As I watch she hunkers down to examine something. She wipes the dirt off and turns it around in her fingers and then, looking pleased, she pops it into the mucky canvas bag slung over her shoulder and continues her leisurely shoreline inspection. I bite my lip against a welt of anger. From the moment of birth, all children do is consume. As early as possible, they need to be educated in their responsibilities, their place in the system. They shouldn't be idling their time away...the word that comes to mind is as antiquated as the city's Victorian legacy...*mudlarking*.

Zooming in on the girl's dishevelled appearance, her jumbled apparel, I realise that she's not a child from one of the regulated accommodations goofing off after all. She's an outsider and she's looking for gifts from the river that she might hope to trade for food. Something soft shifts in my chest and my anger refocuses away from the girl herself to whoever brought her into the world without allowing her access to the resources she needed to live in it. The system isn't perfect, I won't pretend it is, and I understand the reluctance of some to submit to a process that claims ownership of everything you earn, everything you make, consume and throw away. But living like *this* is hardly a viable alternative. Consuming without contributing? That's not fair on everyone else in the community, is it? I'll call it in, but not until I've resolved this thing with Doug.

The girl jumps in surprise as Doug's crawlers burst from the undergrowth in a spindle-legged flurry. They surround her, sniffing and probing at her bag, which she lifts above her head out of their reach. It seems the girl has an eye for treasure. Doug appears then and calls the robots away. Words are exchanged, then laughs. The man's hand gentles the girl's shoulder as the pair head up towards the tenement row but I don't miss the furtive glance around.

Oh, Doug.

Getting out of the car is like opening the door of a furnace. I loosen my tie and top button but keep my uniform jacket on. The mix is heavier on the plastic weave than on the natural fibres but I'm on Ministry business.

From the moment I step through the gate the thicket resists me. The ground is boggy, wiry branches turn me aside, thorns snag my uniform and the air is full of insects. A committed hill walker into late middle age, Mum would have felt at home here.

She'd have had a name for every leaf and flower. I remember only a few. Annoyance at my relative ignorance slides into guilt at the way I palmed her off earlier. I've started to avoid engaging with her because I know how unhappy she is. She'd never in a million years have chosen to live in a place like the Heights, but what alternatives are there?

I press on and eventually find my way to the eerie tenement row which looks like it's in the process of being swallowed up like a Mayan city, so choked is the street with bushes. The sun-bright air carries a sickly sweetness here and bees drowse around the golden gorse flowers. Among them I spot the drone. It's hovering outside the mouth of a close. Within, I can see cracked porcelain tiles, the foot of a staircase. It sparks an unexpected childhood memory of welcome coolness on a hot summer's day that makes me hesitate to go in.

I'm not certain if it's nerves over confronting Doug or guilt over Mum. It doesn't matter. I have a responsibility.

Ducking inside, I find the floor layered with mud and mulch and, as my eyes accustom to the dimness, I see that there are footprints. Loads of them, trailing up the stairs.

On the floor above I find the apartment doors ajar. People can't be trying to live here, can they? But sure enough, in the first apartment I see sleeping bags on the floors. Some of them are occupied. Eyes watch me from the dimness until I beat a retreat. In the second apartment, I stumble across a circle of adults and children, mending clothes the old fashioned way with needles and plastic fibre. The adults look wary. A woman lifts her chin pridefully. 'Can we help you?' she says, and again I retreat.

These outsiders disconcert me. As if their presence is perfectly normal. Like I'm the one who's in the wrong for being here. I can hear some of them following on behind as I climb the next flight of stairs. 'Doug?' I try to put authority into my voice but the old sandstone slaps my words back at me. Again, louder, 'Doug!'

On the next floor I find an apartment turned into a store for junk. Heaps of ancient bits and bobs, presumably dug up by the mudlarks, sorted by type. Piles of plastic. Cairns of buttons and coins. Jumbles of watches, phones, gaming devices from bygone ages. Things that might contain atomic treasures—processors, screens and batteries. All gifted by the river. I snort and barge across the landing, blundering into a kitchen. It's dim because someone has fitted a board over the window, reducing the light to a rectangular halo. My jaw drops. The room is full of tanks. The tanks contain some sort of solution in which scavenged junk is immersed, and in each one there is evidence of...activity. Stuff floating on top or sunk to the bottom.

I know what I'm looking at, I just refuse to believe it. It's an amateur reclamation set-up. Jury-rigged in the kitchen of a disgusting, abandoned tenement. A mockery of the ingenuity and graft our city's had to resort to just to keep going.

Retreating from the kitchen I stumble into another room and find a trio of folding tables containing assorted equipment. The gear has got that coarse, tapioca-coloured finish that speaks of the most basic level of printing and there are wires running everywhere. Among the clutter I spot a boxy chamber that could function as an incubator. Nearby, a screen patched to a phone displays web pages of DNA sequences.

'You've got to be kidding me,' I breathe, astonished equally by the ingenuity on display and the audacity of it all. Outsiders biohacking our technology for their own ends. 'You can't...'

'Who says we can't?' I whirl to face to the woman who spoke to me downstairs. Her arms are folded belligerently and her features are set in a pugnacious scrunch, but that's not the first thing I notice about her. The first thing I notice is how careworn she is. The greying of the hair tucked over her ears. The paper-thin complexion. The shadows around eyes that hold more fear than anger. It's a look that I pretend not to see in the mirror every morning, and it throws me off far more than a stand-up argument would have.

'Lynda?' And now Doug's there too. 'Can we talk in private?'

I follow him meekly up the last flight of stairs to the top floor of the tenement. Here, the flats have been converted to hydroponic gardens. Thick profusions of leaves, stunningly green under the lights. 'I don't understand this,' I tell him. 'Any of it. Why are they doing this?'

Doug's footering with a roller blind. It's warped and soiled with damp and mould but he manages to raise it, letting the sunlight in and giving us a view of the secret country that lies between this forgotten street and the great, brown river. 'What don't you understand, Lynda?' he says quietly. 'People using the resources to hand to survive? Surely, you of all people understand that.'

'But they're not—' I stammer. 'They should be—'

'In the system?' He sighs. 'Why? They have everything they need out here, doing things their own way. They're doing okay. What's wrong with that?'

'What's *wrong* with it?' I stare at him. The question is ridiculous, but for a second I'm not sure of the answer myself. 'How many families are living here, all cramped together?' I say, eventually retreating to the party line. 'In the Heights they'd each have clean, warm, efficient apartments. In the system they'd have jobs.'

'Menial, state-provided jobs, rewarding them with an evening's electricity and just enough calories to get out of bed in the morning and go and do it all again? They reckon they're happier working for themselves.'

'Profiting from what they drag out of the mud you mean?'

'That's only a small part of it. Most of what they reclaim is recycled here to make substrate for printable stuff like solar panels for the electricity or plastic fibre weave for clothes.' Doug rubs his chin. 'Admittedly, the rare earth elements they don't have so much use for.'

'And that's where you come in?'

He shrugs. 'I feed it into the system and pass the value back to them in credit or in kind, that's all.'

'Out of the goodness of your heart, I suppose?' I scoff.

'You find it that difficult to believe?' He shakes his head. 'I get a slice for...hush money, I suppose you could call it.' Then he looks me meaningfully in the eye. 'But most of that I spend trying to help people inside the system. The ones who find it hardest to adjust.'

Even though right is on my side, I'm the one who looks away first. Outside, a breeze has sprung up, enough to riffle through the bushes and raise a chop on the brown water. Clouds have edged into view too. It'll be muggy later but there'll be rain. Glasgow has changed a lot in the last century but it wouldn't be Glasgow without rain. For the first time in an awful long time, I wonder what it is we're trying so desperately to preserve.

'So?' Doug says and I can feel the weight of anticipation in the question to come. Not just his, but that of every lost soul who's found their way to this place. Holding their breath as they crowd the stairs, listening in. 'What are you going to do?'

And the answer to that, I realise with no little shock, is that I don't know. I came here with thirty years of conviction behind me. But witnessing this wild land and the people that have chosen to live in it, my thoughts feel scattered like petals and I'm not sure where they are going to land.

The system works. It is as efficient as it can be. All of its resources are consumed and remediated, again and again, and again. Including its most important components. Its people. There can be no room in the system for this kind of freedom, can there?

What I *should* do is file a report. Bring the Ministry departments down on these outsiders until they beg to be allowed to join the system, and have Doug sanctioned for abetting them too. Down in the street, the bees float and gyre around the bushes.

I am as surprised by the next word I speak as Doug is. 'Nothing.'

'Seriously?'

'I won't turn them in. Things can carry on as normal, but...' My turn to look at him. 'You stop taking a cut. That's not fair.'

He looks like he's about to argue but in the end he just nods. 'And you don't want anything for yourself?'

All this time I've been watching the young girl from before. She's out at the river margin again, poking around with her stick. Her hair is flying in the breeze and she's smiling. Free. Happy. Then on some impulse she looks up and somehow, even at this distance, she sees me watching and waves. By instinct I wave back and the moment is an echo of a memory of a different time. And perhaps an echo of a now that could have been. A simple gesture between mother and child that costs nothing and means everything. A glint of treasure.

'I want Mum to come and live here,' I say.

And I feel renewed.

A Note on the Science
by Louise Horsfall

Certain bacteria and fungi have evolved to survive in environments which have otherwise toxically-high concentrations of dissolved metals. A number of cellular mechanisms enable them to do this, one of which is the transformation of the more toxic dissolved metal ions into less-toxic, nano-sized, clusters of solid particles. If these microorganisms were to be combined with others that are capable of bioleaching—a somewhat opposing process where bacteria and fungi dissolve metals to release them from whatever material they are part of—it would give rise to a full metal recycling system. We are still in the early stages of exploring these ideas, but scientists are identifying and optimising, through synthetic biology, the genetic elements involved in bacterial metal processing with an aim of increasing nanoparticle production rates, metal selectivity, control of nanoparticle size, etc. While the small size of the resultant metal nanoparticles makes them excellent candidates for use as catalysts for various chemical reactions, further properties, often unrelated to the original material, emerge from their nanosize and allow them to be used in a much wider range of applications.

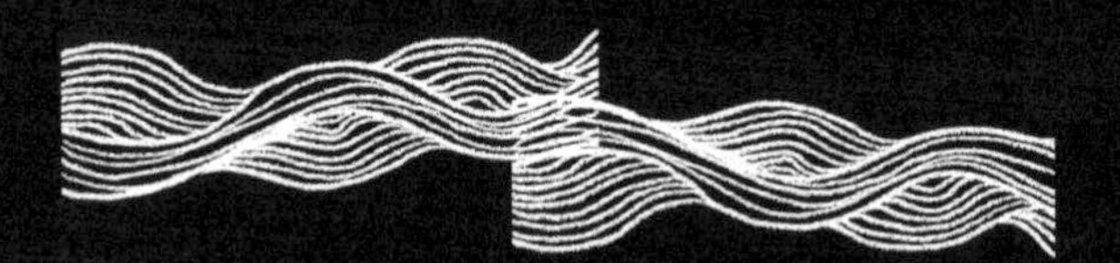

Neil Williamson lives in Glasgow, Scotland. His short stories have been collected in *The Ephemera* and *Secret Language*, both shortlisted for the British Fantasy Award. His other books include *Nova Scotia: New Scottish Speculative Fiction* (with Andrew J Wilson), a finalist for the World Fantasy Award, and *The Moon King*, a finalist for the BSFA British Fantasy Robert Holdstock awards.

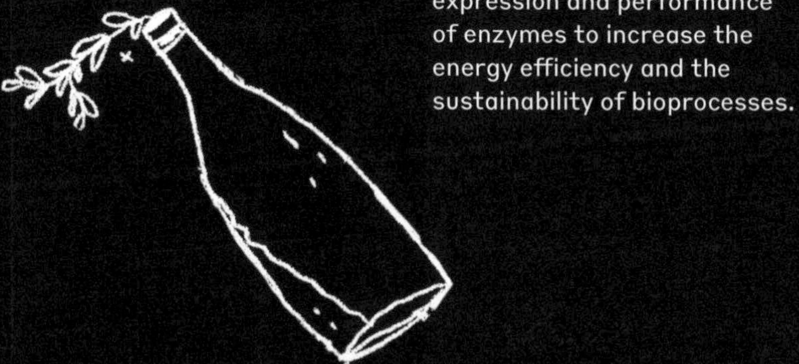

Professor Louise Horsfall is Chair of Sustainable Biotechnology and Programme Director of the MSc in Synthetic Biology and Biotechnology at the University of Edinburgh. She holds an EPSRC fellowship and is the currently elected co-chair of the Bioengineering and Bioprocessing section of the European Federation of Biotechnology. Her research focuses on the application of microbial synthetic biology to move us towards a more sustainable, circular economy. Current research projects include the bioremediation of waste, water and land, employing techniques and tools provided by synthetic biology to increase the value of metals recovered. Collaborative research with industry is focused on improving both the expression and performance of enzymes to increase the energy efficiency and the sustainability of bioprocesses.

Shoreline of Infinity

Shoreline of Infinity is a science fiction and fantasy focused publisher and events host based in Edinburgh, Scotland.

As well as a range of science fiction related publications Shoreline of Infinity also publishes a quarterly science fiction magazine featuring new short stories, poetry, artwork, reviews and articles.

Writers we've published include: Iain M. Banks, Jane Yolen, Nalo Hopkison, Charles Stross, Eric Brown, Ken MacLeod, Ada Palmer, Gary Gibson, Jeannette Ng, Adam Roberts, Jo Walton, Bo Bolander, Tim Majors, Pippa Goldschmidt, Zen Cho, Chri Beckett. We're equally proud of all the new writers we've published.

Shoreline of Infinity Science Fiction Magazine received British Fantasy Society Award 2018 for best magazine/periodical.

Shoreline of Infinity also hosts Event Horizon – a monthly live science fiction cabaret in Edinburgh.

To find out more, visit the website:

www.shorelineofinfinity.com

and follow us on Twitter: @shoreinf

Lightning Source UK Ltd.
Milton Keynes UK
UKHW021840190522
403254UK00008B/1280